Lecture Notes in Energy

Volume 25

For further volumes:
http://www.springer.com/series/8874

Lecture Notes in Energy (LNE) is a series that reports on new developments in the study of energy: from science and engineering to the analysis of energy policy. The series' scope includes but is not limited to, renewable and green energy, nuclear, fossil fuels and carbon capture, energy systems, energy storage and harvesting, batteries and fuel cells, power systems, energy efficiency, energy in buildings, energy policy, as well as energy-related topics in economics, management and transportation. Books published in LNE are original and timely and bridge between advanced textbooks and the forefront of research. Readers of LNE include postgraduate students and non-specialist researchers wishing to gain an accessible introduction to a field of research as well as professionals and researchers with a need for an up-to-date reference book on a well-defined topic. The series publishes single and multi-authored volumes as well as advanced textbooks.

Paula Fernández González
Manuel Landajo · Mª José Presno

The Driving Forces of Change in Environmental Indicators

An Analysis Based on Divisia Index
Decomposition Techniques

 Springer

Paula Fernández González
Manuel Landajo
Mª José Presno
University of Oviedo
Oviedo
Spain

Additional material to this book can be downloaded from http://extras.springer.com

ISSN 2195-1284   ISSN 2195-1292   (electronic)
ISBN 978-3-319-07505-1   ISBN 978-3-319-07506-8   (eBook)
DOI 10.1007/978-3-319-07506-8
Springer Cham Heidelberg New York Dordrecht London

Library of Congress Control Number: 2014940512

Printed on acid-free paper

Springer is part of Springer Science+Business Media (www.springer.com)

To our families

Contents

Introduction and Plan of the Book

The concept of sustainable development leads to a form of progress that aims to satisfy the needs of present generations without compromising the ability of the future ones to meet their own requirements. Major international actors such as the European Union are seeking to establish long-term policies, enabling a style of development that is sustainable from environmental, economic, and social standpoints, with a view to improving the welfare and living conditions of both the present and the future generations.

Historical data witness a strong correlation between energy availability, economic activity, and improvement in living standards and social welfare. Energy is a basic input in the generation of wealth (Nathwani et al. 1992), so the power sector plays a critical role. Any measures adopted on that area must be compatible with the basic principles of competitiveness, security of supply, reduction of energy dependency and vulnerability, and environmental protection.

After the oil crisis of the 1970s, most developed countries have seen a significant shift toward less energy-intensive industries. The numerous factors that might explain that change include: (a) the growing preference of consumers for services and low-energy-intensive materials and products (Williams et al. 1987), (b) the emergence of new, improved materials (Gardner and Robinson 1993) and innovative product designs that attempt to reduce the need for energy to obtain the finished product (Ross et al. 1987), (c) changes in trade patterns (such as the penetration of basic materials from other countries into national markets), (d) the price of energy (Gardner and Elkhafif 1998), and even (e) economic growth itself, since energy-intensive industries are generally more sensitive to investment in infrastructures and equipment. In this work we will address the above issues, focusing our attention on both environmental and energy problems.

In the Kyoto Protocol (1997), successor to the United Nations Framework Convention on Climate Change (1992), industrialized countries commited themselves to reduce their emissions of several greenhouse gases having global warming effects. In particular, the total emissions from developed countries should be reduced during the 2008–2012 period by at least 5 % below 1990 levels. To achieve its objectives, the Protocol proposes a number of ways, including (i) establishment or strengthening of national policies to reduce emissions (by increasing energy efficiency, promoting sustainable forms of agriculture and developing renewable energy sources, among others) and (ii) cooperation with other contracting parties (including exchange of experiences and information, coordination of national policies through emission permits, joint implementation, and mechanisms of clean development). Subsequent agreements signed by the EU—like the so-called Objective 20/20/20 (2008)—have posed a greater commitment to the goals of reducing energy consumption, improving energy efficiency and reducing greenhouse gases.

One of the five goals set up by the European Union in 2011, to be achieved by 2020, is fulfilment of several objectives concerning energy and climate change, for which each Member State shall formulate its own plans and actions. One of the pillars of the European growth strategy is meeting those green objectives. The interactions between environmental, social, and economic phenomena are clearly stated. Therefore, development of proper indicators capable of collecting those interactions becomes of great interest in order to design energy and environmental policies. In this regard, we will focus our attention on decomposition analysis, this being a set of techniques that enable quantification and analysis of the determinant factors influencing changes over time in the main energy and environmental aggregates.

The objective of our study is threefold. First, we analyze several Divisia-index-based decomposition techniques, with the ultimate purpose of quantifying the specific factors in which the variation in a given magnitude may be decomposed. Then we propose and formally analyze a new, exhaustive decomposition technique based on natural spline interpolation, which aims at providing a useful addition to the tool kit of researchers in the field. Finally, we will exploit all the above techniques in order to identify, measure, and analyze the impact of the specific factors that influence some key environmental and energy aggregates, such as energy intensity (defined as the amount of energy used per unit of output) and greenhouse gas emissions. In return, this analysis will provide a number of useful guidelines to the design and implementation of policies to keep those aggregates under control. Energy efficiency (usually proxied by energy intensity) has become crucial to the economic and social development of a country, also being a central component in any environmental program (Jollands and Aulakh 1996). Reduction in greenhouse gas emissions, among which carbon dioxide is included, is one of the main objectives of the international community. An analysis of this environmental aggregate seems of great interest and could lead to conclusions that help establish bottom lines for environmental performance.

Part I of the book, comprising Chaps. 1 and 2, is theoretical in nature and addresses decomposition techniques from a methodological perspective. Part II, including Chaps. 3 and 4, is mainly empirical in nature, reporting various applications of the theory to the energy and environmental fields. In the last section some conclusions are collected.

Chapter 1 reviews the literature on techniques to decompose variations in a generic indicator, presenting an overview of methodologies relying on classical Divisia-type indices. Both multiplicative and additive approaches are considered. Some specific decomposition methods are outlined in the chapter, with implementation details provided in Part II of the book (Chaps. 3 and 4).

In Chap. 2, a new decomposition approach that we denominate *the splines method* is presented. This technique relies on reconstruction of the time paths of the components of the aggregate by using classical polynomial spline interpolation. Some limit theorems are formally derived that ensure convergence, both deterministic and stochastic, to the theoretical quantities of the continuous time Divisia theory. The analysis in the chapter focuses on the problem of decomposing the variation of a ratio (energy intensity) in two factors (namely, the structural and intensity effects), with an application of the method to that specific case being reported in Chap. 3. However, the scope of the theory is far more general, and includes decompositions where an arbitrary number of factors are considered (for instance, a three factor decomposition of a different aggregate is addressed in Chap. 4). In essence, the method may be readily extended to the decomposition of the variation of a generic aggregate, provided that suitable differentiability conditions hold.

Part II opens with Chap. 3. Following the so-called *energy intensity* approach (Ang 1994), we apply several (both parametric and nonparametric) multiplicative decomposition methods to quantify the factors that contributed to the change in aggregate energy intensity in the European economy between years 1995 and 2010. A graphical analysis of the evolution of oil prices in that period is also included in order to assist us in drawing conclusions. Several classical Divisia methods (based on Laspeyres, Marshall-Edgeworth and adaptive weights), as well as refined LMDI and the method of splines, are used to implement the study. A detailed analysis of results is carried out, including a comparison study of the findings provided by the various approaches.

Chapter 4 aims to additively decompose the change in greenhouse gas emissions in the EU15 countries between years 1990 and 2002. For that purpose, in addition to the LMDI and splines methods studied in previous chapters, we introduce two alternative approaches. These are the methods proposed by Sun (1998) and the so-called *path-based approach* (PB; Fernández Vázquez 2004; Fernández Vázquez and Fernández González 2008). Both techniques require a priori specification of parametric models for the time paths followed by the factors and deliver exhaustive decompositions. Sun's method relies on the principle of "jointly created and equally distributed," whereas the PB method provides a more general and flexible approach that exploits statistical information potentially available for intermediate periods in order to determine the trajectories of the factors. Changes in greenhouse gas emissions in the EU15 are analyzed by using these four methods, under several information availability scenarios (namely, annual, triennial, and "periodwise" information). Our analysis extends previous results of Fernández Vázquez and Fernández González (2008).

A summary of the main theoretical and empirical results of the study is collected in the Conclusions section. Broadly speaking, the EU as a whole has achieved a reduction in both greenhouse gas emissions and energy intensity. That drop has been achieved through innovation, adaptation to more efficient techniques, and particularly the increasing use of higher quality energies. As for the future, we strongly recommend an increase of all these efforts in order to fight climate change.

References

Ang BW (1994) Decomposition of industrial energy consumption: the energy intensity approach. Energy Econ 16(3):163–174

Fernández Vázquez E (2004) The use of entropy econometrics in decomposing structural change. Thesis, Universidad de Oviedo, Spain

Fernández Vázquez E, Fernández González P (2008) An extension to Sun's decomposition methodology: the path based approach. Energy Econ 30(3):1020–1036

Gardner DT, Robinson JB (1993) To what end? A conceptual framework for the analysis of energy use. Energy Stud Rev 5(1):1–14

Gardner DT, Elkhafif MAT (1998) Understanding industrial energy use: structural and energy intensity changes in Ontario industry. Energy Econ 20(1):29–41

Jollands N, Aulakh HS (1996) Energy use patterns and energy efficiency trends: the case of energy intensity analysis in New Zealand In: IPENZ annual conference 1996, proceedings of engineering, providing the foundations for society, vol 2, pp 95–100 (see online at: http://www.ema.org.nz/papers/96eupeet.htm)

Nathwani JS, Siddall E, Lind NC (1992) Energy for 300 years. Institute for Risk Research, University of Waterloo, Ontario

Ross M, Larson ED, Williams RH (1987) Energy demand and material flows in the economy. Energy 12(10–11):1111–1120

United Nations (1997) Kyoto protocol to the united nations framework convention to the climate change. United Nations

United Nations (1992) United nations framework convention on climate change. United Nations

Williams RH, Larson ED, Ross MH (1987) Materials, affluence and industrial energy use. Annu Rev Energy 12:99–144

Chapter 1
Literature Review and Methodology

1.1 Introduction

In order to analyse the historical changes in economic, environmental, socio-economic and energy indicators, it is useful to identify, separate and evaluate the macroeconomic forces that contribute to those changes. Basically, the literature records four paradigms that may be used in order to decompose the change experienced by an indicator. These are (a) econometric analysis, (b) analysis based on aggregate data, (c) index-based analysis (*Index Decomposition Analysis*, or IDA), and (d) structural analysis (*Structural Decomposition Analysis*, or SDA).

Econometric decomposition analysis has the potential drawback that it is usually limited to the estimation of first-level effects, by using independent variables from the second level (e.g., Panayotou 1997; de Bruyn et al. 1998). However, the other three techniques allow researchers to separate the various components of a given identity in order to analyse, from a comparative static perspective, to what extent the variation of those components may have contributed to the change in the selected indicator. Among the latter three types of techniques, decomposition analysis of aggregate data is the simplest, since it considers aggregate variables and does not require specific information (e.g., Kaya Identity; Kaya 1990). By contrast, Structural Decomposition Analysis and Index Decomposition Analysis enable use of data from sectoral disaggregation. Given the goal in our study, we will focus on these two techniques. More specifically, we will build on IDA, as it is the most common approach in environmental and energy studies, as well as providing a number of advantages over SDA. In particular, IDA allows both additive and multiplicative decompositions, as well as enabling decompositions for any kind of aggregates (value, ratio, elasticity), and requiring less initial information in multi-country studies.

Within the general framework of Index Decomposition Analysis we will focus on a specific methodology that relies on the Divisia index. That index is commonly used in the field of environmental decompositions. Authors as Richter (1966) and Hulten (1973) have demonstrated that the Divisia index has many nice

P. Fernández González et al., *The Driving Forces of Change in Environmental Indicators*, Lecture Notes in Energy 25, DOI: 10.1007/978-3-319-07506-8_1, © Springer International Publishing Switzerland 2014

properties—including proportionality, homogeneity and invariance—and that, provided it verifies the path independence requirement, it is assured to be equal or superior to any other index.

1.2 Literature Review

Since the 1980s, numerous contributions relying on Divisia-type indices have been published. From an empirical standpoint, works by Jenne and Cattell (1983), Reitler et al. (1987), Boyd et al. (1988), Chen and Rose (1989), Park et al. (1991), Greening et al. (1997), Gardner and Elkhafif (1998), Nag and Parikh (2000), Zhang (2003), Lee and Oh (2006), Hatzigeorgiou et al. (2008), Ma and Stern (2008), Wood (2009), Sahu and Narayanan (2010) and Shahiduzzaman and Alam (2013) have exploited several Divisia-index-based decomposition techniques.

From the methodological perspective, authors such as Hulten (1987), Boyd et al. (1987), Liu et al. (1992), Ang (1994, 1995a, 2005), Ang and Lee (1994), Ang and Choi (1997), Sun (1998), Sun and Ang (2000), Albrecht et al. (2002), Fernández Vázquez and Fernández González (2008) and Choi and Ang (2012) have been concerned with creating a conceptual framework that provides a theoretical foundation to these methods and validates their empirical applications.

In particular, Boyd et al. (1987) analysed the methodologies applied by several authors—relying on time series of energy consumption, production and industrial inputs—and proposed an alternative methodology based on the class of indices derived by Divisia (1925). In the same work they included a list of studies where the trends of energy demand in the industry are analysed, identifying (a) the impact of real energy intensity as a result of improved energy efficiency (defined as the amount of energy used per dollar or tonne of steel produced goods), and (b) the impact of sectoral change, which is a consequence of both the change in the industrial production structure and the displacement of production from sectors more intensive in energy to others that are less energy-intensive.

Later, Liu et al. (1992) proposed the so-called *energy consumption approach*, which is based on two general parametric Divisia methods, exploiting the additive decomposition of changes in aggregate energy consumption into the contributions of changes of three factors, namely production, production structure and sectoral energy intensities. Given the practical impossibility of obtaining continuous time information on the evolution of the relevant variables, Liu et al. (1992) transform the continuous problem into a discrete one, deriving the two general parametric methods we have mentioned. In accordance with the several weights assigned to the magnitudes at each instant, it is possible to obtain several particular methods that may lead to partially different results and conclusions in empirical analysis. This is known as the *problem of non-uniqueness*.

In addition, the application of the above methods conveys the potential drawback of the existence of a residual term that makes that the change in the aggregate does not generally match the sum (or product) of its components. In principle, it seems desirable that there is no deviation of the sum (or product) of the components with respect to the aggregate. If this is the case, the decomposition is said to be *exact* (also denominated *complete, perfect or exhaustive*). In this sense, a useful criterion for the selection of a decomposition method is based on the magnitude of its residual term. A large deviation implies that a significant amount of the observed variation is not explained by any of the components considered, which may adversely affect the purpose of the decomposition. However, a relatively small deviation is a necessary but not sufficient condition for good performance, as it may be the case that the decomposition has large errors in each of its components, whereas the residual term is small as a consequence of aggregation of errors having opposite signs. So, although there is no guarantee on the goodness of a given decomposition, in principle, methods leading to smaller residual terms will be more interesting.

Ang (1994) proposes the so-called *energy intensity approach*, to additively and multiplicatively decompose variations in the ratio between energy consumption and production, in terms of the contributions of the structural and intensity effects. Ang (1995a) proposes the *energy elasticity approach*, which enables estimation of the energy coefficients. That method may be more suitable in cases where it is required to obtain projections for energy demand.

Ang and Choi (1997) proposed the so-called *logarithmic mean Divisia index* (LMDI) method. LMDI is a 'refined', nonparametric approach that relies on Divisia indices with logarithmic mean weights and produces exact decompositions.

Sun (1998) formulates a decomposition technique based on the 'jointly created, equally distributed' principle, whereby each factor is assigned its individual, *ceteris paribus* impact, plus a fraction of the jointly created impact (i.e. the interaction term), which is uniformly distributed among all factors.

Sun's method is nonparametric and exhaustive.[1] It has been improved by Fernández Vázquez (2004) within input–output studies, and subsequently adapted to index-based decomposition analysis by Fernández Vázquez and Fernández González (2008). Its scope now includes applications ranging from environment and energy to technology and productivity.

Recently, Choi and Ang (2012) and Fernández González et al. (2013) have extended the methodology of Divisia-index-based decompositions to energy studies, deriving two new indicators (so-called *structural change index* and *real energy intensity index*) that quantify the individual contributions of each attribute to the percentage change in the factors.

---

[1] Sun's original contribution only considers additive decompositions.

1.3 Divisia-Index-Based Methodology

1.3.1 Structural Decomposition Versus Index-Based Decomposition Techniques

These two approaches[2] have developed independently. While the IDA literature has concentrated on issues such as in-depth analysis of the implications of index number theory and specification of decompositions, the SDA approach has been more focused in distinguishing a large number of effects and determinant factors. The SDA approach enables distinction of a broad spectrum of technological and final demand effects, as well as a number of indirect demand effects that are not likely to be analysed within the IDA environment. However, given its milder information requirements, IDA remains a heavily used tool (e.g., in international studies and time series analysis).

In addition, the IDA literature has developed decompositions for the three main kinds of indicators (absolute variation, ratio, elasticity), while (excepting the case of studies on labour productivity; e.g., Dietzenbacher et al. 2000) the SDA approach often restricts its application to changes in the absolute variation of the indicator. As an additional advantage, the IDA literature has produced both additive and multiplicative decompositions, while SDA typically focuses on the additive form (which is much easier to interpret by non-experts) and rarely applies the multiplicative one.

Given all the above advantages of IDA over the SDA approach, and bearing in mind our goal of study, we will focus on the IDA methodology, also known as 'index number analysis' or simply 'energy decomposition', due to its wide application in environmental studies.

The IDA approach essentially entails the application of economic index number theory to decompose the change in an aggregate indicator into a predefined set of factors. From a methodological perspective, this problem is analogous to the classical index number problem,[3] characterized by Irving Fisher (1922) and well-known in the fields of Economics and Statistics.

1.3.2 Decomposition of the Change in an Indicator

Let V be an aggregate having n factors which contribute to its change over time. For each factor r categories are considered, and we will denote as x_{ij} the value of

---

[2] SDA is the technique most frequently used in decomposition studies based on input–output tables, and therefore it is sometimes known as *input–output decomposition analysis*.

[3] In order to be acceptable, a class of economic indices must preserve all information and be mathematically consistent. Diewert (1980, 1990) derives a number of important developments in economic index number theory.

factor i for category j, where $i = 1,...,n$, $j = 1,...,r$. By definition, the following decomposition of V holds:

$$V = \sum_{j=1}^{r} V_j \tag{1.1}$$

where, for each category j, we have

$$V_j = x_{1j}\, x_{2j} \ldots x_{nj} \tag{1.2}$$

Our interest focuses on the study of the specific factors involved in the change experienced by aggregate V when passing from period 0 ($V^0 = \sum_{j=1}^{r} x_{1j}^0 x_{2j}^0 \ldots x_{nj}^0$) to period T ($V^T = \sum_{j=1}^{r} x_{1j}^T x_{2j}^T \ldots x_{nj}^T$). That change, denominated *total effect*, may be defined at least for three types of indicators: absolute variation, ratio and elasticity. The type of aggregate will depend on the purpose of the study, thus leading to different specific formulations.

1.3.3 Types of Decomposition. Additive and Multiplicative Models

These are the leading decomposition schemes. The additive form decomposes the difference of the indicator between 0 and T (namely, $D_{\text{tot}} = V_T - V_0$) as the sum of the number of effects considered (D_{ieffect}), namely:

$$D_{\text{tot}} = \sum_{i=1}^{n} D_{\text{ieffect}} + D_{\text{rsd}} = \Delta x_{1\text{effect}} + \Delta x_{2\text{effect}} + \cdots + \Delta x_{n\text{effect}} + D_{\text{rsd}} \tag{1.3}$$

The multiplicative form decomposes the relative growth of an indicator between periods 0 and T ($D_{\text{tot}} = V_T/V_0$) in the product of the number of effects considered (R_{ieffect}):

$$R_{\text{tot}} = R_{1\text{effect}}\, R_{2\text{effect}} \ldots R_{n\text{effect}}\, R_{\text{rsd}} \tag{1.4}$$

In the above two expressions, if the decomposition is not exact, a residual term appears (D_{rsd} and R_{rsd}, respectively).

1.3.4 Decomposition Techniques Based on Divisia Indices

In the IDA literature, Liu et al. (1992) extended the Divisia-index-based methodology to the environmental field. Three groups of methods—so-called 'conventional', 'adaptive weights' and 'refined' techniques—have been subsequently developed upon the basis of that work. As noted above, the problem of non-uniqueness arises because we have different methods and alternative approaches available that may lead to a potential mismatch of results.

The Divisia index is a weighted sum of growth rates, with weights corresponding to the contributions of the various components to the total value. Assuming that $\{x_1(t), x_2(t), ..., x_n(t)\}$ is the observation vector to be indexed, with $\{y_1(t), y_2(t), ..., y_n(t)\}$ being the associated price vector, the continuous time version of the Divisia index (Hulten 1973) is defined as:

$$D(\Gamma) = \exp\left(\int_\Gamma \varphi \, d\alpha(t)\right) = \exp\left(\int \left(\sum_{i=1}^n \frac{y_i(t)\, x_i(t)}{\sum_{j=1}^n y_j(t)\, x_j(t)} \frac{dx_i(t)}{x_i(t)}\right)\right) \quad (1.5)$$

where φ is the vector of prices normalized by the value shares, $\alpha(t)$ represents the path of the x's over time interval $[0, T]$, and Γ is the curve described by $\alpha(t)$, $0 \le t \le T$.

The Divisia index has a number of properties—including proportionality, homogeneity and invariance—that are desirable for an economic indicator (Richter 1966). However, a major drawback is its dependence on the path of integration.[4] Provided that this problem may be solved, the Divisia index would conserve, up to the normalization, all the information in the problem and would be, according to Richter (1966), equal or superior to any other index.

Hulten (1973) demonstrates that the Divisia integral is path independent in $S \subset \Re^n$ if and only if there exists an economic aggregate associated with each point $\{x_1, x_2, ..., x_n\} \in S$. This poses a direct relationship between the major disadvantage of Divisia indices—the problem of path dependence—and the theory of aggregation. In any event, that condition is necessary but not sufficient.

Upon results obtained by Richter (1966), Hulten (1973) also derives necessary and sufficient conditions for path independence of the Divisia integral. In particular, it is ensured that $D(\Gamma) = D(\Gamma_1)D(\Gamma_2)...D(\Gamma_t)$, with $\Gamma_1, \Gamma_2,..., \Gamma_t$ being the paths followed by the variable under analysis and Γ being the union of those paths, if and only if

---

[4] In this regard, a line integral is said to be independent of its path in an open region S in $\Re^n$ if (for φ continuous on S, with x and y being two points in S):

$$\int_{\Gamma(x,y)} \varphi \, d\alpha(t) = \int_{\Gamma_1(x,y)} \varphi \, d\beta(t)$$

for all paths $\alpha(t)$ and $\beta(t)$ describing curves Γ and Γ_1 such that $\Gamma \subset S$, $\Gamma_1 \subset S$, $\alpha \subset \Gamma$, and $\beta \subset \Gamma_1$.

(a) There exists an aggregate defined on S.
(b) The aggregate is linearly homogeneous.
(c) There exists and observable price normal at each point in S, unique up to a scalar multiplication.

In practice, given that economic data are not available in continuous time form, a discrete version of the Divisia index is appropriate. The percent change of a discrete Divisia-type index may be defined as follows:

$$\ln(D_t) - \ln(D_{t-1}) = \sum_{i=1}^{n} \left[\rho_i V_{i,t} + \pi_i V_{i,t-1}\right] \left[\ln\left(x_{i,t}\right) - \ln\left(x_{i,t-1}\right)\right] \tag{1.6}$$

where $\rho_i + \pi_i = 1$, $V_{i,t} = \dfrac{y_{i,t} x_{i,t}}{\sum_{j=1}^{n} y_{j,t} x_{j,t}}$; $i = 1, 2, ..., n$.

In his *Index Number Theory,* Fisher (1922) compares results and properties of hundreds of indices, concluding that it is mathematically impossible to find a unique index that simultaneously satisfies all the desirable properties. Many indices have been proposed for use on several decomposition techniques, especially in the IDA approach. Below we detail some of those most commonly appearing in the literature.

1.3.5 Parametric Divisia Methods: Some Classical Methods

Given a certain level of disaggregation, we can express the V indicator as follows:

$$V = \sum_{j=1}^{r} x_{1j} x_{2j} \ldots x_{nj} \tag{1.7}$$

(For the sake of brevity, we will omit the time dependence of the trajectories and write x_{ij} instead of the more rigorous form $x_{ij}(t)$.)

Differentiating Eq. (1.7) with respect to time t we obtain:

$$V' = \sum_{j=1}^{r} x'_{1j} x_{2j} \ldots x_{nj} + \sum_{j=1}^{r} x_{1j} x'_{2j} \ldots x_{nj} + \cdots + \sum_{j=1}^{r} x_{1j} x_{2j} \ldots x'_{nj} \tag{1.8}$$

where x'_{ij} denotes the derivative of x_{ij} with respect to time.

Integrating both sides of (1.8) between times 0 and T we obtain (in the additive decomposition case):

$$\Delta V_0^T = D_{\text{tot}} = \int_0^T \sum_{j=1}^r x'_{1j} x_{2j} \ldots x_{nj} \, dt + \int_0^T \sum_{j=1}^r x_{1j} x'_{2j} \ldots x_{nj} dt + \cdots +$$

$$\int_0^T \sum_{j=1}^r x_{1j} x_{2j} \ldots x'_{nj} dt \tag{1.9}$$

Dividing by the V indicator and integrating both sides from 0 to T, the following multiplicative decomposition is obtained:

$$\frac{V_T}{V_0} = R_{\text{tot}} = \exp\left(\int_0^T \sum_{j=1}^r x'_{1j} x_{2j} \ldots x_{nj} \frac{V_j(t)}{V(t)} \, dt \right)$$

$$\exp\left(\int_0^T \sum_{j=1}^r x_{1j} x'_{2j} \ldots x_{nj} \frac{V_j(t)}{V(t)} \, dt \right) \ldots \exp\left(\int_0^T \sum_{j=1}^r x_{1j} x_{2j} \ldots x'_{nj} \frac{V_j(t)}{V(t)} \, dt \right) \tag{1.10}$$

$$\exp\left(\int_0^T \sum_{j=1}^r x_{1j} x_{2j} \ldots x_{nj} \frac{V'_j(t)}{V(t)} \, dt \right)$$

In practice, the above path integrals are usually approximated by discretization. From Eqs. (1.9) and (1.10), and under certain conditions of linear homogeneity in the aggregate magnitudes (relatively easy to assume for sufficiently small time intervals), Liu et al. (1992) derived the general *parametric Divisia methods* 1 and 2 (hereafter denoted as PDM1 and PDM2, respectively). Both methods differ only in their weighting structures: while the former applies the Montgomery-Vartia index, the latter opts for the Vartia-Sato index.

We will consider the following particularizations of the above general parametric Divisia methods:

(1) *Laspeyres Parametric Divisia Method* 1 (LAS-PDM1). It is a special case of PDM1, with all weights equal to 0. Its denomination as an Laspeyres-type method stems from the fact that it only takes into account the values of the parameters in the initial period.

(2) *Paasche Parametric Divisia Method* 1 (PAA-PDM1). It constitutes a particular case of PDM1 with all weights equal to unity. It is regarded as a 'Paasche' method because it only considers quantities in the final period. This method, in both multiplicative and additive forms, has been applied by Hoekstra and Bergh (2003).

(3) *Marshall-Edgeworth Parametric Divisia Method* 1 (AVE-PDM1). It is a special case of PDM1 that uses an arithmetic mean of the Laspeyres and Paasche indexes. All weights have value 0.5. In its multiplicative form, this method was proposed by Boyd et al. (1987).

(4) *Laspeyres Parametric Divisia Method* 2 (LAS-PDM2). It is a special case of PDM2 with weights equal to 0. It is similar to the method proposed by Howarth et al. (1991) and Jenne and Cattell (1983) in its additive form, while Hankinson and Rhys (1983) and Park (1992) derived the multiplicative version.

(5) *Paasche Parametric Divisia Method* 2 (PAA-PDM2). It constitutes a special case of PDM2, with unit weights.

(6) *Marshall-Edgeworth Parametric Divisia Method* 2 (AVE-PDM2). It is a special case of PDM2 wherein the weights take the value 0.5. It is equivalent to the proposal by Reitler et al. (1987).

(7) *Adaptive Weights Parametric Divisia Method* (AWT-PDM).[5] In this case, the results of the decomposition are the same regardless of the parametric Divisia method that is used, as the weights are obtained by equalizing the expressions of Parametric Divisia Methods 1 and 2. We will omit the specific reference to PDM1 or PDM2, and simply use the PDM denomination. This method was proposed by Liu et al. (1992).

As above commented, a proper selection of the decomposition method becomes crucial in practice, as different methods can lead to contradictory conclusions for the same data set. This potential disparity in results arises from the differences in assumptions incorporated in the formulation of each method. As noted by Ang and Lee (1994), it is generally preferable to use PDM1 if the relationship between the variable to decompose and the predefined factors has approximately logarithmic form, while PDM2 should be preferred if the relationship is approximately linear.

Concerning the weights, the main advantage of methods based on simple averaging is that their results are symmetrical with respect to time, since a fixed base year is not specified but an arithmetic mean of the values of the relevant variables in periods 0 and T is used instead. Boyd et al. (1988) and Reitler et al. (1987) regarded as advantageous characteristics of these methods their symmetrical nature and the fact that the weights are not referred to any specific year.

Laspeyres-type methods usually convey larger residual terms than the remaining methods. As an advantage, those methods are capable to isolate the individual effect of each factor, decomposing the change in the intensity of an aggregate indicator in terms of the changes in the relevant variables between the initial and final periods (respectively, 0 and T), with all other variables kept constant at their values in the base year. Park (1992) considers this a logical form of decomposition, as in his opinion it would correspond to the economic concept of marginal effect of a variable, so decomposition results can be easily interpreted.

The adaptive weights method is regarded by Liu et al. (1992) as unique in the sense that, unlike the above methods, the analyst is not forced to arbitrarily assign parameter values. On the other hand, a potential drawback of the method is its relative computational complexity.

---

[5] The term 'adaptive' refers to the fact that the parameters are not fixed in advance. Instead, they are determined by the levels of the magnitudes in periods 0 and T.

Finally, the choice of weights should be consistent with the objective of the study. For instance, if our objective is to observe the behaviour pattern of a give indicator, weights such as 0 and 0.5 may be the best choice. Conversely, if our aim is prediction, values greater than 0.5 or even 1 may be more useful. By assigning value 1 to the weight, both decomposition and prediction may be carried out with respect to the same reference period.

1.3.6 Nonparametric Divisia Method. The LMDI Approach

When the factors to be identified in the decomposition of a given aggregate do not undergo drastic changes in the period of study, published studies suggest that conventional Divisia methods behave relatively well in terms of residuals. However, when significant changes are observed in the data—which is quite frequent in data with a high disaggregation level or in decompositions of the intensity of a particular energy source—the conventional Divisia method may not be suitable as it generally entails a relatively large residual term. In those situations, the refined Divisia index method (or *LMDI*) would be preferable as it ensures the absence of residual term in the decompositions.

The refined Divisia index method was introduced by Ang and Choi (1997), who proposed a new kind of logarithmic mean weighting that leads to exact decompositions. Their idea may be connected with the work of Sato (1976) who, in search for an ideal index, proposed the following symmetric weight function, obtained from the logarithmic mean[6] of x and y defined as

$$L(x,y) = \frac{(y - x)}{\ln(y/x)} \tag{1.11}$$

With x and y being positive numbers, $x \neq y$. It may be shown that

$$L(x,x) = \lim_{y \to x} L(x,y) = x \tag{1.12}$$

By replacing x and y in the above equation by $w_{j,0}$ and $w_{j,T}$, respectively, we obtain:

$$L(w_{j,0}, w_{j,T}) = \frac{(w_{j,T} - w_{j,0})}{\ln(w_{j,T}/w_{j,0})} \tag{1.13}$$

---

[6] Törnqvist et al. (1985) argue that the above logarithmic function was originally proposed by Törnqvist in (1935), though it is in their 1985 work where it was eventually specified that:

(a) x and y should be positive numbers,

(b) if $x \neq y$ the range of function $L(x,y)$ should be $\left((xy)^{1/2}, (x+y)/2\right)$.

where $w_{j,0} = \frac{x_{j,0}y_{j,0}}{x_0 y_0}$, $w_{j,T} = \frac{x_{j,T}y_{j,T}}{x_T y_T}$,

$x_t = \sum_{j=1}^{r} x_{j,t}$, $y_t = \sum_{j=1}^{r} y_{j,t}$, $t = 0, T$, and r is the number of categories.

However, given the range of the logarithmic mean, in the above formulation the sum of weights is slightly less than unity. This drawback may be solved by a suitable rescaling. The normalized weight function becomes

$$w_j^* = \frac{L(w_{j,0}, w_{j,T})}{\sum\limits_{j=1}^{r} L(w_{j,0}, w_{j,T})}, j = 1, \ldots, r \qquad (1.14)$$

The refined Divisia method is nonparametric, and although requiring a higher number of calculations, it incorporates a great advantage over conventional methods, as it delivers complete decompositions. When variations in the data are drastic along the study period, that advantage becomes crucial. However, authors such as Sun (1998) consider that the residual term is a factor to be clearly spelled out as it collects interesting interactions that may be indicative of promising research avenues.

1.3.7 Decomposition at Several Disaggregation Levels of Information

The disaggregation level defines the number of subcategories to be aggregated. Choosing a specific disaggregation level potentially affects the results and conclusions of the study, so it can be advisable to perform the analysis at several different disaggregation levels. This idea is regarded in the literature as *multilevel decomposition*. Authors like Morovic et al. (1989), Li et al. (1990), Gardner (1993), Sahu and Narayan (2010) and Fernández González et al. (2014a) have applied that approach.

Considering only two disaggregation levels, with level I corresponding to the various categories (or groups) and level II to subcategories (or sectors), the decomposition in each respective level is given by:

$$D_{tot} = D1_{1effect} + D1_{2effect} + \cdots + D1_{neffect} + D1_{rsd} \qquad (1.15)$$

$$D_{tot} = D2_{1effect} + D2_{2effect} + \cdots + D2_{neffect} + D2_{rsd} \qquad (1.16)$$

if the decomposition is additive, and

$$R_{tot} = R1_{1effect} R1_{2effect} \ldots R1_{neffect} R1_{rsd} \qquad (1.17)$$

$$R_{tot} = R2_{1effect}R2_{2effect} \ldots R2_{neffect}R2_{rsd} \qquad (1.18)$$

in the multiplicative case.

Relying on Ang (1995b), it is possible to decompose the effect of an arbitrary factor i in the upper disaggregation level ($D2_{\text{ieffect}}$ or $R2_{\text{ieffect}}$, effects at the sector level) into the contribution of that factor in a lower level ($D1_{\text{ieffect}}$ or $R1_{\text{ieffect}}$, effects at the group level) and its contribution when going in the bottom-up direction ($D12_{\text{ieffect}}$ or $R12_{\text{ieffect}}$, subgroup effects). Namely,

$$D2_{\text{ieffect}} = D1_{\text{ieffect}} + D12_{\text{ieffect}} , \ i = 1, \ldots, n \qquad (1.19)$$

in the additive case, and

$$R2_{\text{ieffect}} = R1_{\text{ieffect}} \, R12_{\text{ieffect}} , \ i = 1, \ldots, n \qquad (1.20)$$

when the decomposition is multiplicative.

By generalizing the above result, and considering h disaggregation levels, the total effect may be expressed as:

$$\begin{aligned} D_{\text{tot}} = {}&(D1_{1\text{effect}} + D12_{1\text{effect}} + \cdots + D\{h-1\}h_{1\text{effect}}) + \cdots + \\ &(D1_{n\text{effect}} + D12_{n\text{effect}} + \cdots + D\{h-1\}\,h_{n\text{effect}}) + Dh_{\text{rsd}} \end{aligned} \qquad (1.21)$$

for the additive decomposition, and

$$\begin{aligned} R_{\text{tot}} = {}&(R1_{1\text{effect}} \, R12_{1\text{effect}} \ldots R\{h-1\}h_{1\text{effect}}) \cdots \\ &(R1_{n\text{effect}} \, R12_{n\text{effect}} \ldots R\{h-1\}h_{n\text{effect}}) \, Rh_{\text{rsd}} \end{aligned} \qquad (1.22)$$

if the decomposition is multiplicative.

1.3.8 Time Series Decomposition

For any given time interval, decomposition techniques may be applied either statically (so-called *periodwise decomposition)* or dynamically (*time series decomposition*). In the former case only the initial and the final periods are considered, whereas in time series decompositions the change in the aggregate is expressed as the cumulative effect of the changes occurring in intermediate periods. Let $(C_{\text{tot}})_{0,T}$ be the total cumulative change in aggregate V between periods 0 and T, $(C_{\text{ieffect}})_{0,T}$ the estimated cumulative effect for factor i, and $(C_{\text{rsd}})_{0,T}$ the cumulative residual term between 0 and T. The time series decomposition is carried out by considering any consecutive periods $(t-1, t)$, where $t = 1, \ldots, T$. We have:

(a) Additive decomposition:

$$(C_{tot})_{0,T} = (D_{tot})_{0,1} + (D_{tot})_{1,2} + \cdots + (D_{tot})_{T-1,T} \tag{1.23}$$

$$(C_{1effect})_{0,T} = (D_{1effect})_{0,1} + (D_{1effect})_{1,2} + \cdots + (D_{1effect})_{T-1,T} \tag{1.24}$$

$$(C_{2effect})_{0,T} = (D_{2effect})_{0,1} + (D_{2effect})_{1,2} + \cdots + (D_{2effect})_{T-1,T} \tag{1.25}$$

$$\cdots$$

$$(C_{rsd})_{0,T} = (D_{rsd})_{0,1} + (D_{rsd})_{1,2} + \ldots + (D_{rsd})_{T-1,T} \tag{1.26}$$

(b) Multiplicative decomposition:

$$(C_{tot})_{0,T} = (R_{tot})_{0,1} (R_{tot})_{1,2} \ldots (R_{tot})_{T-1,T} \tag{1.27}$$

$$(C_{1effect})_{0,T} = (R_{1effect})_{0,1} (R_{1effect})_{1,2} \ldots (R_{1effect})_{T-1,T} \tag{1.28}$$

$$(C_{2effect})_{0,T} = (R_{2effect})_{0,1} (R_{2effect})_{1,2} \ldots (R_{2effect})_{T-1,T} \tag{1.29}$$

$$\cdots$$

$$(C_{rsd})_{0,T} = (R_{rsd})_{0,1} (R_{rsd})_{1,2} \ldots (R_{rsd})_{T-1,T} \tag{1.30}$$

As evident from the above expressions, dynamic analysis takes into account all the intermediate periods, unlike periodwise decompositions where only the initial (0) and final (T) periods are considered. Obviously, the latter requires less information and more reduced efforts in data collection and calculation of results. These advantages come at the cost of completely ignoring the growth patterns of the variables in intermediate years. That may be a serious limitation that explains why, despite its greater requirements, time series decomposition is generally preferred as it tends to provide results that are more accurate and less method-dependent, since its residual term is usually smaller. Evidently, higher performance is a consequence of the fact that the time path implicitly assumed for the calculation of the integral is interpolated for every two consecutive years, instead of only in periods 0 and T, as occurs in periodwise decompositions.

In any event, periodwise decompositions are appropriate only in cases where availability of data is an issue. This may happen, among others, when the decomposition is carried out at a very high disaggregation level, or in the case of comparative studies between countries. Many decomposition studies on industrial energy consumption apply the periodwise technique instead of time series decomposition. Examples of this type of analysis can be found, among others, in the works of Ang and Skea (1994), Liu et al. (1992) and Park et al. (1991).

References

Albrecht J, Francois D, Schoors K (2002) A Shapley decomposition of carbon emissions without residuals. Energy Policy 30(9):727–736

Ang BW (1994) Decomposition of industrial energy consumption: the energy intensity approach. Energy Econ 16(3):163–174

Ang BW (1995a) Decomposition methodology in industrial energy demand analysis. Energy 20(11):1081–1095

Ang BW (1995b) Multilevel decomposition of industrial energy consumption. Energy Econ 17(1):39–51

Ang BW (2005) The LMDI approach to decomposition analysis: a practical guide. Energy Policy 33(7):867–871

Ang BW, Choi KH (1997) Decomposition of aggregate energy and gas emission intensities for industry: a refined Divisia index method. Energy J 18(3):59–73

Ang BW, Lee SY (1994) Decomposition of industrial energy consumption: some methodological and application issues. Energy Econ 16(2):83–92

Ang BW, Skea JF (1994) Structural change sector, sector disaggregation and electricity consumption in the UK industry. Energy Environ 5(1):1–16

Boyd G, Hanson DA, Sterner T (1988) Decomposition of changes in energy intensity: a comparison of the Divisia index and other methods. Energy Econ 10(4):309–312

Boyd G, McDonald JF, Ross M, Hanson DA (1987) Separating the changing composition of US manufacturing production from energy efficiency improvements: a Divisia index approach. Energy J 8(2):77–96

de Bruyn SM, van den Bergh JC, Opschoor JB (1998) Economic growth and emissions: reconsidering the empirical basis of environmental Kuznets curves. Ecol Econ 25(2):161–175

Chen CY, Rose A (1989) A structural decomposition analysis of changes in energy demand in Taiwan: 1971–1984. Energy J 11(1):127–146

Choi KH, Ang BW (2012) Attribution of changes in Divisia real energy intensity index: an extension to index decomposition analysis. Energy Econ 34(1):171–176

Dietzenbacher E, Hoen A, Los B (2000) Labor productivity in Western Europe 1975–1985: an intercountry, interindustry analysis. J Region Sci 40(3):425–452

Diewert WE (1980) Recent developments in the economic theory of index numbers: capital and the theory of productivity. Am Econ Rev 70(2):260–267

Diewert WE (1990) Price level measurement. North-Holland, Amsterdam

Divisia, F. L. (1925): "Indice monétaire et la théorie de la monnaie," Revue dEconomie Politique, 39, pp. 980-1008

Fernández González P, Landajo M, Presno MJ (2013) The Divisia real energy intensity indices: evolution and attribution of percent changes in 20 European countries from 1995 to 2010. Energy 58(1):340–349

Fernández González P, Landajo M, Presno MJ (2014a) Multilevel LMDI decomposition of changes in aggregate energy consumption. A cross country analysis in the EU-27. Energy Policy 68:576–584

Fernández Vázquez E (2004) The use of entropy econometrics in decomposing structural change. Universidad de Oviedo (Spain), Thesis

Fernández Vázquez E, Fernández González P (2008) An extension to Suńs decomposition methodology: the path based approach. Energy Econ 30(3):1020–1036

Fisher I (1922) The making of index numbers; a study of their varieties, tests, and reliability. Houghton Mifflin, Boston

Gardner DT (1993) Industrial energy use in Ontario from 1962 to 1984. Energy Econ 15(1):25–32

Gardner DT, Elkhafif MAT (1998) Understanding industrial energy use: structural and energy intensity changes in Ontario industry. Energy Econ 20(1):29–41

Greening LA, Davis WB, Schipper L, Khrushch M (1997) Comparison of six decomposition methods: application to aggregate energy intensity for manufacturing in 10 OECD countries. Energy Econ 19(3):375–390

Hankinson GA, Rhys MM (1983) Electricity consumption, electrticity intensity and industrial structure. Energy Econ 5(3):146–152

Hatzigeorgiou E, Polatidis H, Haralambopoulos D (2008) CO_2 emissions in Greece for 1990–2002: a decomposition analysis and comparison of results using the arithmetic mean Divisia index and logarithmic mean Divisia index techniques. Energy 33(3):492–499

Hoekstra R, van der Bergh JCJM (2003) Comparing structural and index decomposition analysis. Energy Econ 25(1):39–64

Howarth RB, Schipper L, Duerr PA, Strøm S (1991) Manufacturing energy use in eight OECD countries. Energy Econ 13(2):135–142

Hulten CR (1973) Divisia index numbers. Econometrica 41(6):1017–1025

Hulten CR (1987) Divisia index. In: Eatwell J et al. (Eds) The new palgrave dictionary of economics. Palgrave Macmillan. (See online at: http://www.dictionaryofeconomics.com/article?id=pde1987_X000612). DOI:10.1057/9780230226203.2399. Palgrave Macmillan. Accessed 14 Sept 2012

Jenne C, Cattell R (1983) Structural change and energy efficiency in industry. Energy Econ 5(2):114–123

Kaya Y (1990) Impact of carbondioxide emission control on GDP growth: interpretation and proposed scenarios," IPCC Energy and Industry Subgroup, Response Strategies Working Group, Paris.

Lee K, Oh W (2006) Analysis of CO_2 emissions in APEC countries: a time-series and a cross-sectional decomposition using the log mean Divisia method. Energy Policy 34(17):2779–2787

Li JW, Shrestha RM, Foell WK (1990) Structural change and energy use: the case of the manufacturing sector in Taiwan. Energy Econ 12(2):109–115

Liu XQ, Ang BW, Ong HL (1992) The application of the Divisia index to the decomposition of changes in industrial energy consumption. Energy J 13(4):161–177

Ma C, Stern DI (2008) China's changing energy intensity trend: a decomposition analysis. Energy Econ 30(3):1037–1053

Morović T, Gerritse G, Jaeckel G, Jochem E, Mannsbart W, Poppke H, Witt B (1989) Energy conservation indicators II. Springer, Berlin

Nag B, Parikh J (2000) Indicators of carbon emission intensity from commercial energy use in India. Energy Econ 22(4):441–461

Panayotou T (1997) Demystifying the environmental Kuznets curve misleading us? The case of CO_2 emissions. Environ Dev Econ 2(4):465–484

Park SH (1992) Decomposition of industrial energy consumption—an alternative method. Energy Econ 14(4):265–270

Park SH, Dissmann B, Nam KY (1991) A cross-country decomposition analysis of manufacturing energy consumption. Energy 18(2):93–100

Reitler W, Rudolph M, Schaefer M (1987) Analysis of the factors influencing energy consumption in industry: a revised method. Energy Econ 9(3):145–148

Richter MK (1966) Invariance axioms and economic indexes. Econometrica 34(4):739–755

Sahu SK, Narayanan K (2010) Decomposition of industrial energy consumption in Indian manufacturing: the energy intensity approach. J Environ Manag Tour Assoc Sustain Educ, Res Sci 1:22–38

Sato K (1976) The ideal log-change index number. Rev Econ Stat 58(2):223–228

Shahiduzzaman Md, Alam K (2013) Changes in energy efficiency in Australia: a decomposition of aggregate energy intensity using Logarithmic Mean Divisia approach. Energy Policy 56:341–351

Sun JW (1998) Changes in energy consumption and energy intensity: a complete decomposition model. Energy Econ 20(19):85–100

Sun JW, Ang BW (2000) Some properties of an exact decomposition model. Energy 25(12):1177–1188

Törnqvist L (1935) A memorandum concerning the calculation of Bank of Finland consumption price index. Bank of Finland, Helsinki

Törnqvist L, Vartia P, Varita Y (1985) How should relative changes be measured? Am Stat 39(1):43–46

Wood R (2009) Structural decomposition analysis of Australia's greenhouse gas emissions. Energy Policy 37(11):4943–4948

Zhang Z (2003) Why did the energy intensity fall in China's industrial sector in the 1990s? The relative importance of structural change and intensity change. Energy Econ 25(6):625–638

Chapter 2
Mathematical and Statistical Properties of Decomposition Techniques. The Splines Method

2.1 Introduction

As discussed in Chap. 1, the theoretical Divisia index is calculated upon the basis of the continuous time paths of the observed variables. However, only a finite number of discrete observations is available in practice. As seen in the above chapter, two basic strategies may be applied in order to alleviate that discrepancy between continuous theory and discrete data. Most authors have opted for discretizing theory. This is the spirit of methods such as general PDM1 and PDM2 discussed in Chap. 1. Alternatively, a few contributions in the literature have taken the opposite direction, i.e., generating approximate continuous time paths that more properly adapt to the theory. This is the basic idea of some techniques as the path-based method (Fernández Vázquez and Fernández González 2008).

In this chapter we propose a new continuous time decomposition method that is based on spline interpolation. Our analysis relies on classical results of the mathematical theories of spline interpolation and approximation of functions (e.g., Powell 1981, Chap. 23), as well as on certain stochastic analogues of these results. We begin by studying some mathematical properties of additive and multiplicative decompositions. For the sake of brevity we shall focus on the specific problem of decomposing the variation of a ratio (namely, the energy intensity ratio) in two factors (respectively, structural and intensity effects), although our theoretical results are valid in more general cases, as to be detailed below.

Our proposal may be regarded as nonparametric, since no functional form is assumed for the (deterministic or stochastic) paths to be reconstructed. We shall only impose the requirement of convergence of the approximate, spline-interpolation-based decompositions to their theoretical, continuous time analogues.

Our analysis will follow this sequence: first, upon the basis of a finite set of discrete observations of the relevant variables (namely, production levels and

P. Fernández González et al., *The Driving Forces of Change in Environmental Indicators*, Lecture Notes in Energy 25, DOI: 10.1007/978-3-319-07506-8_2, © Springer International Publishing Switzerland 2014

energy consumptions), a reconstruction of their continuous time trajectories is generated.[1]

Secondly, approximations to the quantities of interest (namely, the intensity change for the period under study and its components) in Divisia-based decomposition analysis are constructed by plugging the interpolated trajectories into the relevant path integrals. Then, convergence—as the sampling is performed on an increasingly finer time mesh—towards the exact (continuous time) decomposition of energy intensity is derived. Finally, the analysis is extended to the stochastic field, by the expedient of assuming that the time paths are generated by continuous time stochastic processes having appropriate regularity properties.

2.2 Path Reconstruction Through Interpolation. The Splines Method

We shall consider classical polynomial splines, with a single variable t (the time index), where $-\infty \leq a \leq t \leq b \leq \infty$ for arbitrary values a and b. Following the usual definition (e.g., Powell 1981, Chap. 3, page 29), a piecewise polynomial function $Q(t)$ is called a spline of degree K in $[a, b]$ if $Q(t)$ is a polynomial of degree K in each section and has continuous derivatives up to order $K - 1$ in $[a, b]$. Formally we will say that $Q(t)$ belongs to the function space $C^{K-1}[a, b]$ of all functions with continuous derivatives up to order $K - 1$ in $[a, b]$. In particular, we are interested in the approximation capabilities of splines on interval $[0,1]$. In that specific case, every spline $Q(t)$ of degree K is characterized by a set $N_n = \{t_1, \ldots, t_n\}$ of points called *knots*, such that $-\infty \leq a < 0 = t_1 < t_2 < \cdots < t_n = 1 < b \leq \infty$ and the spline may be expressed in the following form:

$$Q(t) = \sum_{j=0}^{K} c_j t^j + \sum_{i=2}^{n-1} d_i (t - t_i)_+^K \tag{2.1}$$

where c_j and d_j are constants and $()_+^K$ denotes the truncated power of degree K, i.e., $(z)_+^K = [\max(0, z)]^K$.

Splines are very flexible structures that allow function interpolation in a way that preserves a number of interesting features, such as monotonicity and convexity of the interpolated functions (e.g., DeVore and Lorentz 1993, Chap. 13).

---

[1] Path reconstruction can be accomplished through a number of interpolation techniques. Two basic requirements are derived from the characteristics of the above theoretical decomposition: (a) the interpolants must be able to approximate the relevant *time paths and their derivatives* up to first order, and (b) the proposed methods should lead to exact decompositions of the variation in energy intensity. These two requirements, as well as computational simplicity, are fulfilled by the splines method.

Within the many spline-based approximation and interpolation techniques, those based on so-called *natural splines* have particularly interesting capabilities. Let $W^m[a, b]$ be the Sobolev space consisting of all functions having continuous derivatives in $[a, b]$ up to order $m - 1$ (i.e., the set functions belonging to the function space $C^{m-1}[a, b]$ and having square integrable mth order derivatives in $[a, b]$, with $m = 1, 2, \ldots$; e.g., Adams 1975, Chap. 3). Natural splines were proposed by Schoenberg (1964) as a solution to the following variational problem (we slightly adapt Schoenberg's general formulation to our problem): find a function $\hat{f}$ belonging to $W^m[a, b]$ which solves the following minimization problem

$$\min_{\hat{f} \in W^m[a, b]} \int_a^b \left(D^m \hat{f}(t) \right)^2 dt \tag{2.2}$$

subject to the set of conditions $\hat{f}(t_i) = f(t_i)$, where $i = 1, \ldots, n$ and $-\infty \leq a < 0 = t_1 < t_2 < \cdots < t_n = 1 < b \leq \infty$. Classical results (e.g., Powell 1981, Chap. 23, Theorems 23.1–23.2) show that, provided that the number of observations is $n \geq m$, the above variational problem has a unique solution which is a natural spline of order m, i.e., a spline of degree $2m - 1$ and continuous derivatives up to order $2m - 2$.

An especially interesting feature of natural splines of order m is that they are able to uniformly approximate arbitrary smooth functions and their derivatives up to order $m - 1$ on compact sets. Proposition 1 below formally states this property (again this is a particularization, adapted to our setting, of a general property of natural splines; e.g., Wahba 1990; also see Schultz 1973): we shall focus on the problem of approximating in interval $0 \leq t \leq 1$ an arbitrary function f belonging to Sobolev space $W^m[a, b]$, with $-\infty \leq a < 0 \leq t \leq 1 < b \leq \infty$, upon the basis of a sample of n observations (i.e., n points in the graph of the function).[2] In the remainder of the chapter we will denote by $D^\alpha f$ the αth order derivative of f, with $D^0 f = f$.

Proposition 1 (Wahba 1990, pp. viii–ix) *Let f be a function belonging to $W^m[a, b]$. Then, for some constant $L < \infty$, the natural spline of interpolation $\hat{f}$ (with order m and degree $2m - 1$) that interpolates f at $N_n = \{t_i = (i - 1)/(n - 1), i = 1, \ldots, n\}$, satisfies*

$$\max_{0 \leq \alpha \leq m-1} \sup_{0 \leq t \leq 1} \left| D^\alpha \hat{f}(t) - D^\alpha f(t) \right| \leq L(n - 1)^{-(m-\alpha)} \tag{2.3}$$

---

[2] For notational simplicity we will consider data evenly spaced in time, although the results of this chapter are valid for unevenly spaced observations, under the condition that the maximum distance between any two consecutive observations converges to zero as n goes to infinity.

Fig. 2.1 Function
$f(t) = (t + 0.02)^{1/2}$ (*dotted line*) and spline interpolant (*solid line*) with $n = 4$ observations (represented as *asterisks*)

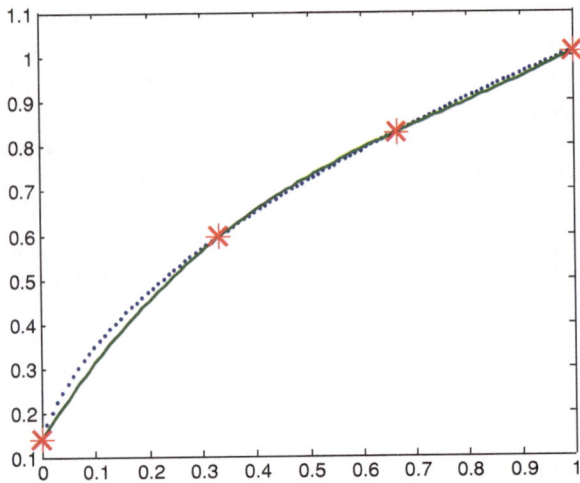

Proposition 1 ensures that, under very general conditions, if the function to be reconstructed is smooth enough, the natural spline interpolant and its derivatives up to order $m - 1$ converge uniformly to f and its respective derivatives, as the graph of the function is more densely sampled. Our derivations in this chapter rely heavily on this property.

As a simple illustration, Fig. 2.1 displays a reconstruction of $f(t) = (t + 0.02)^{1/2}$, $0 \leq t \leq 1$, on the basis of $D_n = \{(t_i, f(t_i)), t_i = (i - 1)/(n - 1), i = 1, 2, \ldots, n\}$, using only $n = 4$ observations.

As shown in Fig. 2.1, the deviation between the function and its natural (cubic) spline interpolant is barely noticeable.

2.3 Mathematical Properties. Convergence

The above smooth approximation properties of splines can be applied to path approximation in the general Divisia problem. As to be shown below, decompositions based on spline interpolation converge to the theoretical (continuous time) solution of the Divisia problem. We shall focus on the representative case of the decomposition of the variation of energy intensity in a given period. For simplicity we consider the unit interval, $0 \leq t \leq 1$, with t being the time index (any other finite interval may be chosen instead).

We shall assume that the economy is composed of r sectors ($j = 1, \ldots, r$), and the following notation will be used:

$e_j(t)$: Instantaneous energy consumption in sector j evaluated at time t,

$y_j(t)$: Instantaneous production of sector j at time t,

$e(t) = \sum_{j=1}^{r} e_j(t)$: Total energy consumption at t,

$y(t) = \sum_{j=1}^{r} y_j(t)$: Total production at t.

2.3.1 Additive Decomposition

Aggregate energy intensity at time t is defined as usual:

$$I(t) = \frac{e(t)}{y(t)} \qquad (2.4)$$

As $e(t) = \sum_{j=1}^{r} e_j(t)$, the following decomposition is readily obtained:

$$I(t) = \sum_{j=1}^{r} \frac{e_j(t)}{y(t)} = \sum_{j=1}^{r} \frac{e_j(t)}{y_j(t)} \frac{y_j(t)}{y(t)} = \sum_{j=1}^{r} I_j(t) \cdot S_j(t) \qquad (2.5)$$

where $I_j(t) = e_j(t)/y_j(t)$ is the energy intensity in sector j and $S_j(t) = y_j(t)/y(t)$ is the share of sector j in the total production at time t.

Under Assumptions 1 and 2 below, the intensity function $I(t)$ has a continuous first order derivative in $[0,1]$ admitting the following decomposition:

$$D^1 I(t) = \sum_{j=1}^{r} D^1 I_j(t) \cdot S_j(t) + \sum_{j=1}^{r} I_j(t) \cdot D^1 S_j(t) \qquad (2.6)$$

The above expression directly results in the following additive decomposition of the intensity variation:

$$I_1 - I_0 = TE = IE + SE \qquad (2.7)$$

where $I_0 = I(0)$, $I_1 = I(1)$, TE is the total effect (or intensity change), $IE = \int_0^1 \left(\sum_{j=1}^{r} D^1 I_j(t) S_j(t) \right) dt$ is the intensity effect, and $SE = \int_0^1 \left(\sum_{j=1}^{r} I_j(t) D^1 S_j(t) \right) dt$ is the structural effect, all of them referred to the accumulation period between $t = 0$ and $t = 1$.

More generally, the variation of total intensity between 0 and t, denoted by $TE(t)$, with $0 \le t \le 1$, may be decomposed as $TE(t) = IE(t) + SE(t)$, with $IE(t) = \int_0^t \sum_{j=1}^{r} D^1 I_j(u) S_j(u) \, du$ being the intensity effect accumulated up to t and

$SE(t) = \int_0^t \sum_{j=1}^r D^1 S_j(u) I_j(u)\,du$ being the structural effect accumulated along the same period, so that evidently $TE(1) = I_1 - I_0$.

We shall consider natural cubic spline interpolants, denoted by $\hat{e}_j(t)$ and $\hat{y}_j(t)$, respectively, for the time paths $e_j(t)$ and $y_j(t)$, $j = 1, \ldots, r$, $0 \leq t \leq 1$. Spline interpolants for total energy consumption $e(t)$ and total production $y(t)$ are readily obtained upon the basis of $e_j(t)$ and $y_j(t)$, respectively, as $\hat{e}(t) = \sum_{j=1}^r \hat{e}_j(t)$ and $\hat{y}(t) = \sum_{j=1}^r \hat{y}_j(t)$, which evidently coincide with the natural spline interpolants for $e(t)$ and $y(t)$, and under the conditions of Proposition 1 below will converge uniformly to $e(t)$ and $y(t)$, respectively, and their derivatives up to order 1. Plug-in interpolants for the intensities and production shares are defined directly on the basis of $\hat{e}_j(t)$ and $\hat{y}_j(t)$. In particular:

$\hat{I}_j(t) = \hat{e}_j(t)/\hat{y}_j(t)$: interpolant for energy intensity in sector j,

$\hat{I}(t) = \hat{e}(t)/\hat{y}(t)$: interpolant for aggregate energy intensity,

$\hat{S}_j(t) = \hat{y}_j(t)/\hat{y}(t)$: interpolant for the production share of sector j.

Evidently, the above three interpolants are no longer splines, although they inherit most approximation capabilities of splines $\hat{e}_j(t)$ and $\hat{y}_j(t)$, which suffices for our purposes.

The above structures interpolate the discrete set of observations of the relevant variables (consumptions, productions, intensities, shares) and, as to be shown below, they also provide suitable approximations to the continuous time paths of these variables and their first derivatives.

We shall consider the following plug-in approximants to functions $IE(t)$, $SE(t)$ and $TE(t)$, respectively, $\hat{I}E(t) = \int_0^t \sum_{j=1}^r D^1 \hat{I}_j(u)\hat{S}_j(u)\,du$, $\hat{S}E(t) = \int_0^t \sum_{j=1}^r \hat{I}_j(u) D^1 \hat{S}_j(u)\,du$ and $\hat{T}E(t) = \hat{I}E(t) + \hat{S}E(t)$, with $0 \leq t \leq 1$. Evidently, $\hat{I}E = \hat{I}E(1)$, $\hat{S}E = \hat{S}E(1)$ and $\hat{T}E = \hat{T}E(1)$ will be, respectively, plug-in approximants for IE, SE and TE, corresponding to the whole accumulation period.

To ensure that the above interpolants satisfy Proposition 1 we will impose the following regularity conditions on the paths we want to reconstruct:

Assumption 1 For each $j = 1, \ldots, r$: (i) sector consumption $e_j(t)$ has continuous derivatives up to order 1 in [0,1], with $\int_0^1 (D^2 e_j(t))^2 dt \leq c < \infty$, and (ii) $e_j(t) \geq m > 0$.

Assumption 2 For each $j = 1, \ldots, r$: (i) sector production $y_j(t)$ has continuous derivatives up to order 1 in [0,1], with $\int_0^1 (D^2 y_j(t))^2 dt \leq c < \infty$, and (ii) $y_j(t) \geq m > 0$.

The following proposition shows that the above approximations converge to the theoretical, continuous time effects as the number of observations (n) and its denseness in $[0, 1]$ increase.

Proposition   2   (Convergence   of   the   additive   decomposition) *Let*

$$\hat{TE}_n = \int_0^1 \sum_{j=1}^r \left[D^1 \hat{I}_j(t) \hat{S}_j(t) + \hat{I}_j(t) D^1 \hat{S}_j(t) \right] dt \text{ be the plug-in approximant to TE}$$

generated through natural spline interpolation (of order 1 and degree 3) of the time paths, applied to the set of observations $D_n = \{(t_i, e_1(t_i), \ldots, e_r(t_i),$ $y_1(t_i), \ldots, y_r(t_i)),\ t_i = (i-1)/(n-1);\ i = 1, \ldots, n\}.$ *Then the following holds under Assumptions 1 and 2:*

(a) $\hat{TE}_n \to TE$ *as* $n \to \infty,$ *and in particular*

(b) $\left| \hat{TE}_n - TE \right| \le L_1 (n-1)^{-1}$ *for some* $L_1 < \infty$ *and all n large enough.*   □

An analogue result holds for the plug-in approximations for *IE* and *SE*, and for those of functions $\hat{IE}(t)$, $\hat{SE}(t)$ and $\hat{TE}(t)$ themselves (see the proof of Proposition 2 in Appendix I below).

2.3.2 *Multiplicative Decomposition*

The above ideas are readily extended to the multiplicative case, where the logarithmic total effect is defined as follows:

$$LTE = LTE(1) = ln(I_1/I_0) = \int_0^1 D^1 \ln I(t)\ dt =$$

$$\int_0^1 \sum_{j=1}^r \frac{D^1 I_j(t) S_j(t)}{I(t)}\ dt + \int_0^1 \sum_{j=1}^r \frac{I_j(t) D^1 S_j(t)}{I(t)}\ dt \tag{2.8}$$

The intensity effect accumulated until t is $R_{\text{int}}(t) = \exp(LIE(t))$, with $LIE(t) = \int_0^t \sum_{j=1}^r \frac{D^1 I_j(u) S_j(u)}{I(u)}\ du$ being the logarithmic intensity effect for the $[0, t]$ period. Similarly the structural effect accumulated up to t is $R_{\text{str}}(t) = \exp(LSE(t))$, where $LSE(t) = \int_0^t \sum_{j=1}^r \frac{I_j(t)\ D^1 S_j(t)}{I(t)}\ dt$ is the logarithmic structural effect.

The magnitude to decompose is the total effect accumulated at t, which is simply $R(t) = R_{\text{int}}(t)\ R_{\text{str}}(t)$, with $R(1) = I_1/I_0$ being the ratio of intensities at $t = 1$ and $t = 0$.

We will approximate *LTE* by its spline-based analogue, namely,

$$\hat{LTE}_n = \int_0^1 \sum_{j=1}^r \frac{D^1 \hat{I}_j(t)\, \hat{S}_j(t)}{\hat{I}(t)}\, dt + \int_0^1 \sum_{j=1}^r \frac{\hat{I}_j(t)\, D^1 \hat{S}_j(t)}{\hat{I}(t)}\, dt \qquad (2.9)$$

Analogously to the additive case, the following convergence result holds for (2.9):

Proposition 3 (Convergence of the multiplicative decomposition) *Let $\hat{LTE}_n$ be the plug-in approximant to LTE generated through natural interpolation splines (of order 1 and degree 3) by using the observation set $D_n = \{(t_i, e_1(t_i), \ldots, e_r(t_i), y_1(t_i), \ldots, y_r(t_i)),\ t_i = (i-1)/(n-1);\ i = 1, \ldots, n\}$. Then the following holds under Assumptions 1 and 2:*

(a) *$\hat{LTE}_n \to LTE$ as $n \to \infty$, and in particular*

(b) *$|\hat{LTE}_n - LTE| \le L_1(n-1)^{-1}$ for some $L_1 < \infty$ and all n large enough.*

Therefore, the logarithmic total effect (and also the intensity and structural ones) are uniformly approximated by their natural spline analogues.

2.4 Stochastic Analysis

In this section we will analyze the behaviour of $\hat{TE}_n$ and $\hat{LTE}_n$ in a probabilistic setting where the data used for path reconstruction are generated by a sufficiently regular stochastic process. We will show that, under general conditions, these two approximants are random variables and converge with probability 1 (and therefore in distribution) to TE and LTE (which are also random variables), respectively, as $n \to \infty$.

Let $(\Omega, \mathbf{A}, P)$ be a complete probability space. The set of time paths in $[0, 1]$ will be given by vector function $Z = (z_1, \ldots, z_{2r})$, where $z_j = e_j$, $z_{r+j} = y_j$, $j = 1, \ldots, r$ are the trajectories of energy consumption and production for each of the r sectors. Thus, for each (t, ω) with $t \in [0, 1]$ and $\omega \in \Omega$, we will have $Z(t, \omega) = (e_1(t, \omega), \ldots, e_r(t, \omega), y_1(t, \omega), \ldots, y_r(t, \omega))$, which is a vector of observations at time t. By allowing t to range between 0 and 1, a path vector is obtained. Regarding the set of trajectories, we assume as in the previous section that, for all fixed $\omega \in \Omega$, each component of $Z = Z(t, \omega)$ has continuous derivatives up to order 1 in $[0, 1]$. Thus, we will say that Z is defined on function space $S = \prod_{i=1}^{2r} C^1[0, 1]$, endowed with the metric induced by the norm $||Z|| = \max_{j=1, \ldots, 2r} \max_{0 \le \alpha \le 1} \max_{0 \le t \le 1} |D^\alpha z_j(t)|$. Equipped with this norm, S is a complete separable metric space.[3]

---

[3] See Dudley (1973) for sets of conditions ensuring differentiability, continuity and Lipschitz properties for stochastic processes, both Gaussian and non-Gaussian.

We shall denote by $\mathbf{B}(S)$ the class of Borel sets in S, i.e., the smallest σ-algebra containing all open subsets (in the sense of the metric induced by the norm $||.||$) of S. The ordered pair $(S, \mathbf{B}(S))$ is a probabilizable space. We assume that the stochastic process Z is a random element of S, that is, a measurable mapping of Ω into S. Therefore, for each event B in $\mathbf{B}(S)$, an event A in $\mathbf{A}$ exists such that $Z(A) = B$. Note that the norm $||.||$ is a $\mathbf{B}(S)$-measurable function. The mapping $P'(B) = P(Z^{-1}(B))$, $B \in \mathbf{B}(S)$, defines a probability measure P' induced by Z, and a final probability space $(S, \mathbf{B}(S), P')$. (For simplicity we will also use the same symbol P to denote P', with the proper interpretation to be deduced from the context in each case.)

For each $m > 0$ we will denote by A_m the set of paths $Z = (z_1, \ldots, z_{2r}) \in S$ such that $\min_{j=1,\ldots,2r} \min_{0 \leq t \leq 1} D^1 z_j(t) \geq m$. We will impose the following condition, which is a stochastic analogue of Assumptions 1 and 2 above:

Assumption 1' For every $\omega \in \Omega$: (i) each component of $Z(., \omega)$ belongs to $W^2[a, b]$, and (ii) for some $m > 0$ not depending on $\omega \in \Omega$, it holds $Z(., \omega) \in A_m$.[4]

Under Assumption 1', the total effects TE and LTE, can be expressed as $TE = g(Z)$ and $LTE = h(Z)$, respectively, where $g(.)$ and $h(.)$ are functionals of Z with expressions given by Propositions 2 and 3, respectively. It is readily verified that both g and h are continuous in A_m, that is, for every $Z, Z' \in A_m$, $||Z' - Z|| \to 0$ implies $|g(Z') - g(Z)| \to 0$ and $|h(Z') - h(Z)| \to 0$. (This may be readily deduced by the same procedure used in the proof of Propositions 2 and 3 and Lemmas A.1–A.3 in Appendix I). As Z is $\mathbf{B}(S)$-measurable, continuity of $g(.)$ and $h(.)$ implies that both $TE = g(Z)$ and $LTE = h(Z)$ are random variables (e.g., Billingsley, 1968, Appendix II, p. 222), i.e. measurable functions with respect to the σ-algebra $\mathbf{B}(\Re)$, defined on the real line $\Re$.

We now consider the probabilistic behaviour of $\hat{TE}_n$ and $\hat{LTE}_n$, constructed as described in the previous section. It is easily derived that both $\hat{TE}_n$ and $\hat{LTE}_n$ are random variables under Assumption 1'. For each $\omega \in \Omega$ the sample realizations $\hat{TE}_n(\omega)$ and $\hat{LTE}_n(\omega)$ of $\hat{TE}_n$ and $\hat{LTE}_n$ are standard Riemann integrals so extension to the stochastic field is straightforward.[5]

Proposition 4 *Under Assumption 1':*
(a) For all n large enough $\hat{TE}_n$ and $\hat{LTE}_n$ are $\mathbf{B}(\Re)$-measurable.
(b) As $n \to \infty$, $\hat{TE}_n \to TE$ and $\hat{LTE}_n \to LTE$ with probability 1 (and therefore in distribution).

---

[4] The results in these section are also obtained if Assumption 1' holds almost surely, i.e., for each $\omega \in \Omega$ excepting a set $N \in \mathbf{A}$ with $P(N) = 0$.

[5] The proof of Proposition 4 in Appendix I relies on a version of the Continuous Mapping Theorem that requires all integrals to be defined pointwise, i.e., as standard Riemann integrals for almost each $\omega \in \Omega$. Therefore, use of more general (mean square) stochastic integrals is not sufficient for our purposes. In addition, mean square integration requires more restrictive conditions, such as finite variances (e.g., Tanaka 1996, Chap. 3, page 71), not imposed in Assumption 1' above.

2.5 Concluding Remarks

In this chapter we have established—using as an illustration the problem of decomposing the change in energy intensity—that the exact decompositions based on spline interpolation of the time paths converge to the values derived from the theory of continuous time Divisia indices. These theoretical quantities may be seen as (deterministic and stochastic, respectively) limits of sequences of spline-based approximations. To obtain these results we have relied on some mathematical properties of classical cubic spline interpolation.

The splines method provides a workable alternative to mainstream techniques developed so far in the IDA literature. It incorporates a number of advantages, including (i) its being nonparametric in nature, and (ii) its *objective* (or fully automatic) character, as it does not depend on any parameters to be subjectively chosen by the researcher. In addition, (iii) it is an exhaustive method (in the sense of having zero residual) that naturally verifies the circular property of index numbers, thus allowing so-called time series decompositions. Finally, (iv) the splines method may be applied in multilevel decompositions, provided that time series of sufficient length are available for all the quantities involved in the decomposition analysis.

The results of this chapter can be readily extended to a number of closely related problems. For instance, in Chap. 3 below the splines method is applied to a decomposition of energy intensity under a different number of factors than considered in this chapter, and in Chap. 4 it is used to decompose the variation of an absolute magnitude (namely, greenhouse gas emissions). More generally, the splines method may be applied to the (respectively, additive or multiplicative) decomposition of the variation of the product of any finite number of components or time paths (e.g., Fernández Vázquez and Fernández González 2008), and more generally to decomposing the variation of a differentiable functional of a vector of smooth time paths. The arguments developed in this chapter for the case of energy intensity directly extend to those more general settings.

Appendix I: Mathematical Proofs of Chapter 2

The following two lemmas are required for the proof of Proposition 2.

Lemma A.1 *Under Assumptions 1 and 2, there exists a constant $B < \infty$ such that the following holds for $j = 1, \ldots, r$ and $0 \leq \alpha \leq 1$:*

(a) $\max_{t \in [0,1]} \left| D^\alpha \hat{e}_j(t) - D^\alpha e_j(t) \right| \leq B(n-1)^{-(2-\alpha)}$, and

(b) $\max_{t \in [0,1]} \left| D^\alpha \hat{y}_j(t) - D^\alpha y_j(t) \right| \leq B(n-1)^{-(2-\alpha)}$.

Proof The Lemma is a direct consequence of Proposition 1.

Lemma A.2 *Under Assumptions 1 and 2, there exist constants $B_k < \infty, k = 1, \ldots, 4$ such that, for each $j = 1, \ldots, r$, the following holds for $0 \leq t \leq 1$ and $n > 1 + (2B/m)^{1/2}$, where B is as in Lemma A.1 above and does not depend on t:*

(a) $\left| \hat{I}_j(t) - I_j(t) \right| \leq B_1(n-1)^{-2}$,

(b) $\left| \hat{S}_j(t) - S_j(t) \right| \leq B_2(n-1)^{-2}$,

(c) $\left| D^1 \hat{I}_j(t) - D^1 I_j(t) \right| \leq B_3(n-1)^{-1}$,

(d) $\left| D^1 \hat{S}_j(t) - D^1 S_j(t) \right| \leq B_4(n-1)^{-1}$.

Proof In the proof of this Lemma (and throughout the rest of the Appendix) we will rely on the fact that any continuous function in $[0, 1]$ is bounded in that interval. This implies that, under Assumptions 1(i) and 2(i), there exists a constant $M < \infty$ such that $\max_{0 \leq \alpha \leq 1} \max_{0 \leq t \leq 1} \left| D^\alpha e_j(t) \right| \leq M$ and $\max_{0 \leq \alpha \leq 1} \max_{0 \leq t \leq 1} \left| D^\alpha y_j(t) \right| \leq M$ for each $j = 1, \ldots, r$. The same is true for the aggregate consumption and production functions (respectively, $e(t)$ and $y(t)$).

Let us select an arbitrary point $t \in [0, 1]$. Regarding statement (a), we have: $\left| \hat{I}_j(t) - I_j(t) \right| = \left| \frac{\hat{e}_j(t)}{\hat{y}_j(t)} - \frac{e_j(t)}{y_j(t)} \right| \leq A_I + A_{II}$, where $A_I = \left| \frac{\hat{e}_j(t)}{\hat{y}_j(t)} - \frac{e_j(t)}{\hat{y}_j(t)} \right|$ and $A_{II} = \left| \frac{e_j(t)}{\hat{y}_j(t)} - \frac{e_j(t)}{y_j(t)} \right|$. Let $A_{III} = \hat{y}_j(t) - y_j(t)$. Lemma A.1 ensures both $\left| \hat{e}_j(t) - e_j(t) \right| \leq B(n-1)^{-2}$ and $\left| \hat{y}_j(t) - y_j(t) \right| \leq B(n-1)^{-2}$. Thus, arbitrarily small $\left| \hat{e}_j(t) - e_j(t) \right|$ and $|A_{III}|$ can be obtained for large n. In particular, if $n > 1 + (2B/m)^{1/2}$ we have $A_{III} < m/2$. Therefore, Assumption 2(ii) implies that, for any $n > 1 + (2B/m)^{1/2}$, it holds $\hat{y}_j(t) = y_j(t) + A_{III} \geq m - m/2 = m/2 > 0$, so $A_I = \left| \hat{y}_j(t) \right|^{-1} \left| \hat{e}_j(t) - e_j(t) \right| \leq \frac{B}{m/2}(n-1)^{-2}$.

As for A_{II} we have:

$$A_{II} = \left| e_j(t) \right| \cdot \left| \frac{y_j(t) - \hat{y}_j(t)}{y_j(t)\hat{y}_j(t)} \right|$$

As $\left| e_j(t) \right| \leq M < \infty$ by continuity in $[0, 1]$, Lemma A.1(b) ensures that, for $n > 1 + (2B/m)^{1/2}$, it holds $A_{II} \leq \frac{M}{m^2/2} B(n-1)^{-2}$.

Therefore, $A_I + A_{II} \leq B_1(n-1)^{-2}$ for all n large enough and some finite B_1. As t was arbitrary, uniform convergence is obtained, which completes the proof of statement (a).

Regarding (b) we have, for any $t \in [0, 1]$, $\left| \hat{S}_j(t) - S_j(t) \right| = \left| \frac{\hat{y}_j(t)}{\hat{y}(t)} - \frac{y_j(t)}{y(t)} \right| \leq A_I + A_{II}$, where $A_I = \frac{\left| \hat{y}_j(t) - y_j(t) \right|}{\hat{y}(t)}$ and $A_{II} = \left| y_j(t) \frac{y(t) - \hat{y}(t)}{y(t)\hat{y}(t)} \right|$.

For $n > 1 + (2B/m)^{1/2}$ it holds $\hat{y}(t) = \sum_{j=1}^{r} \hat{y}_j(t) = \sum_{j=1}^{r} y_j(t) + \sum_{j=1}^{r} \left(\hat{y}_j(t) - y_j(t) \right) \geq rm - rm/2 = rm/2$.

Thus $A_I \leq \frac{2}{rm} B(n-1)^{-2}$ and $A_{II} \leq \frac{2M}{r^2 m^2} B(n-1)^{-2}$, so for $n > 1 + (2B/m)^{1/2}$ we have $\left| \hat{S}_j(t) - S_j(t) \right| \leq B_2 (n-1)^{-2}$. Again, as t is arbitrary, convergence is uniform.

Regarding (c), select an arbitrary t in $[0,1]$ and apply the quotient rule for derivatives. We have $\left| D^1 \hat{I}_j(t) - D^1 I_j(t) \right| = |A_I - A_{II}|$, where

$$A_I = \frac{D^1 \hat{e}_j(t) \hat{y}_j(t) - \hat{e}_j(t) D^1 \hat{y}_j(t)}{\left(\hat{y}_j(t) \right)^2}$$

and

$$A_{II} = \frac{D^1 e_j(t) y_j(t) - e_j(t) D^1 y_j(t)}{\left(y_j(t) \right)^2}.$$

By the triangle inequality

$|A_I - A_{II}| \leq |A_I - A_{III}| + |A_{III} - A_{II}|$, where $A_{III} = \frac{D^1 e_j(t) y_j(t) - e_j(t) D^1 y_j(t)}{\left(\hat{y}_j(t) \right)^2}$.

It is directly obtained that, for $n > 1 + (2B/m)^{1/2}$, there exists $B_5 < \infty$ (not depending on n) such that $|A_I - A_{II}| \leq B_5 (n-1)^{-1}$.

Analogously, continuity (and so boundedness in $[0,1]$) of the first derivatives of y_j and e_j implies that, for some $M < \infty$, $\left| D^1 e_j(t) y_j(t) - e_j(t) D^1 y_j(t) \right|$ $\leq \left| D^1 e_j(t) \right| \cdot \left| y_j(t) \right| + \left| e_j(t) \right| \cdot \left| D^1 y_j(t) \right| \leq 2M^2$.

So it follows, for $n > 1 + (2B/m)^{1/2}$ and some $B_6 < \infty$ (not depending on n), $|A_{II} - A_{III}| \leq \frac{4M^2}{m^2} B_6 (n-1)^{-2}$.

Therefore, $|A_I - A_{II}| \leq |A_I - A_{III}| + |A_{III} - A_{II}| \leq B_3 (n-1)^{-1}$ for $n > 1 + (2B/m)^{1/2}$ and some B_3 finite, with convergence being uniform.

As for (d), a similar procedure is applied. We have $\left| D^1 \hat{S}_j(t) - D^1 S_j(t) \right| \leq |A_I - A_{III}| + |A_{III} - A_{II}|$, where $A_I = \frac{D^1 \hat{y}_j(t) \hat{y}(t) - \hat{y}_j(t) D^1 \hat{y}(t)}{(\hat{y}(t))^2}$, $A_{II} = \frac{D^1 y_j(t) y(t) - y_j(t) D^1 y(t)}{(y(t))^2}$ and $A_{III} = \frac{D^1 y_j(t) y(t) - y_j(t) D^1 y(t)}{(\hat{y}(t))^2}$.

Again we obtain that, for $n > 1 + (2B/m)^{1/2}$, there exists $B_6 < \infty$ (not depending on n) so that $|A_I - A_{II}| \leq B_6 (n-1)^{-1}$.

As $\left| D^1 y_j(t) y(t) - y_j(t) D^1 y(t) \right| \leq \left| D^1 y_j(t) \right| \cdot |y(t)| + |y_j(t)| \cdot \left| D^1 y(t) \right| \leq rM^2$ for some $M < \infty$, and since $\left| (y(t))^2 - (\hat{y}(t))^2 \right| = |y(t) + \hat{y}(t)| \cdot |y(t) - \hat{y}(t)| \leq \left(2M + B(n-1)^{-2} \right) B(n-1)^{-2}$, we obtain for $n > 1 + (2B/m)^{1/2}$:

$$|A_{III} - A_{II}| \leq \frac{\left| D^1 y_j(t) y(t) - y_j(t) D^1 y(t) \right| \cdot \left| (y(t))^2 - (\hat{y}(t))^2 \right|}{(y(t))^2 (\hat{y}(t))^2}$$

$$\leq \frac{4rM^2(2M+1)}{r^4 m^4} \left(2M + B(n-1)^{-2} \right) B(n-1)^{-2}$$

Therefore, for each $n > 1 + (2B/m)^{1/2}$ and $0 \leq t \leq 1$, there exists $B_4 < \infty$ (not depending on n or t), such that $|A_I - A_{II}| \leq |A_I - A_{III}| + |A_{III} - A_{II}| \leq B_3(n-1)^{-1}$.

Lemma A.3 Let $D^1 I(t) = \sum_{j=1}^{r} \left(D^1 I_j(t) S_j(t) + I_j(t) D^1 S_j(t) \right)$ and $D^1 \hat{I}(t) = \sum_{j=1}^{r} \left(D^1 \hat{I}_j(t) \hat{S}_j(t) + \hat{I}_j(t) D^1 \hat{S}_j(t) \right)$. Under Assumptions 1 and 2, there exists a constant $B_5 < \infty$ such that, for each $0 \leq t \leq 1$ and $n > 1 + (2B/m)^{1/2}$, with B being as in Lemma A.1, it holds $\left| D^1 \hat{I}_j(t) - D^1 I_j(t) \right| \leq B_5(n-1)^{-1}$.

Proof It straightforwardly derives from Lemma A.2, which establishes that each component of $D^1 \hat{I}(t)$ converges uniformly to its analogue in $D^1 I(t)$. In order to obtain uniform convergence we will rely on the fact that all the components appearing in $D^1 I(t)$ are uniformly bounded in $0 \leq t \leq 1$. In particular:
$$|I_j(t)| = \left| \frac{e_j(t)}{y_j(t)} \right| \leq \frac{M}{m}, \quad |S_j(t)| = \left| \frac{y_j(t)}{y(t)} \right| \leq \frac{M}{rm}, \quad |D^1 I_j(t)| = \frac{|D^1 e_j(t) y_j(t) - e_j(t) D^1 y_j(t)|}{(y_j(t))^2} \leq \frac{2M^2}{m^2} \text{ and}$$
$$|D^1 S_j(t)| = \frac{|D^1 y_j(t) y(t) - y_j(t) D^1 y(t)|}{(y(t))^2} \leq \frac{2M^2}{r^2 m^2}.$$

Similar uniform bounds are obtained for the components of $D^1 \hat{I}(t)$, since according to Lemma A.2, as $n > 1 + (2B/m)^{1/2}$ we have, for each $0 \leq t \leq 1$,
$$|\hat{I}_j(t)| \leq |I_j(t)| + |\hat{I}_j(t) - I_j(t)| \leq \frac{M}{m} + B_1(n-1)^{-2},$$
$$|\hat{S}_j(t)| \leq |S_j(t)| + |\hat{S}_j(t) - S_j(t)| \leq \frac{M}{rm} + B_2(n-1)^{-2},$$
$$|D^1 \hat{I}_j(t)| \leq |D^1 I_j(t)| + |D^1 \hat{I}_j(t) - D^1 I_j(t)| \leq \frac{2M^2}{m^2} + B_3(n-1)^{-1}$$
and
$$|D^1 \hat{S}_j(t)| \leq |D^1 S_j(t)| + |D^1 \hat{S}_j(t) - D^1 S_j(t)| \leq \frac{2M^2}{r^2 m^2} + B_4(n-1)^{-1}.$$

By using the decomposition $D^1 \hat{I}(t) - D^1 I(t) = A_I + A_{II}$, where

$$A_I = \sum_{j=1}^{r} \left(D^1 \hat{I}_j(t) \hat{S}_j(t) + \hat{I}_j(t) D^1 \hat{S}_j(t) \right) - \sum_{j=1}^{r} \left(D^1 \hat{I}_j(t) S_j(t) + \hat{I}_j(t) D^1 S_j(t) \right)$$

and

$$A_{II} = \sum_{j=1}^{r} \left(D^1 \hat{I}_j(t) S_j(t) + \hat{I}_j(t) D^1 S_j(t) \right) - \sum_{j=1}^{r} \left(D^1 I_j(t) S_j(t) + I_j(t) D^1 S_j(t) \right).$$

and applying the above bounds, it is readily obtained that there exists some $B_5 < \infty$ such that, for each t in $[0,1]$ and $n > 1 + (2B/m)^{1/2}$, it holds $\left| D^1 \hat{I}(t) - D^1 I(t) \right| \leq |A_I| + |A_{II}| \leq B_5(n-1)^{-1}$.

Proof of Proposition 2 It is a direct consequence of Lemma A.3. It is readily checked that $D^1 I(t)$ is continuous in $[0,1]$, so it is also bounded in that interval. In particular, for each t in $[0,1]$, it holds $|D^1 I(t)| = \frac{|D^1 e(t) y(t) - e(t) D^1 y(t)|}{(y(t))^2} \leq \frac{2M^2}{m^2}$. An analogous result is obtained for $D^1 \hat{I}(t)$ when $n > 1 + (2B/m)^{1/2}$, as $\hat{y}(t) \geq rm/2 > 0$ in that case, so $D^1 \hat{I}(t)$ is a ratio of continuous functions, with strictly positive denominator (having a positive lower bound not depending on n).

Applying Lemma 3, we obtain

$$\left|\hat{TE}_n - TE\right| = \left|\int_0^1 D^1\hat{I}(t)dt - \int_0^1 D^1 I(t)dt\right| \leq \int_0^1 \left|D^1\hat{I}(t) - D^1 I(t)\right|dt \leq B_5(n-1)^{-1}\int_0^1 dt = B_5(n-1)^{-1}$$

for $n > 1 + (2B/m)^{1/2}$, which proves part *(b)*, and therefore part *(a)*.   □

Proof of Proposition 3

It is a consequence of Lemma A.3. Let us select an arbitrary point t in $[0, 1]$ and apply the decomposition $\left|\frac{D^1\hat{I}(t)}{\hat{I}(t)} - \frac{D^1 I(t)}{I(t)}\right| \leq A_I + A_{II}$, where $A_I = \frac{|D^1\hat{I}(t) - D^1 I(t)|}{\hat{I}(t)}$ and $A_{II} = \left|\frac{D^1 I(t)}{\hat{I}(t)} - \frac{D^1 I(t)}{I(t)}\right|$.

The inequality $\left|D^1\hat{I}(t) - D^1 I(t)\right| \leq B_5(n-1)^{-1}$ was obtained in the proof of Lemma A.3 above, for each t in $[0, 1]$ and $n > 1 + (2B/m)^{1/2}$. In addition, $\frac{m}{M} \leq \frac{1}{I(t)} = \frac{y(t)}{e(t)} \leq \frac{M}{m}$.

It is readily shown, by a procedure analogous to that used in Lemma A.2 *(a)*, that $\left|\hat{I}(t) - I(t)\right| \leq B_6(n-1)^{-2}$ for each t in $[0, 1]$ and $n > 1 + (2B/m)^{1/2}$, with $B_6 < \infty$ not depending on t.

Since $\frac{1}{\hat{I}(t)} = \frac{1}{I(t)} + \frac{I(t)-\hat{I}(t)}{I(t)\hat{I}(t)}$, and as $\left|\hat{I}(t) - I(t)\right| \leq B_6(n-1)^{-2} \leq \frac{m}{2M}$ for all large enough n (for this, it suffices to select $n \geq \sqrt{B_6\frac{2M}{m} - 1}$), we arrive at $\hat{I}(t) = I(t) + \left(\hat{I}(t) - I(t)\right) \geq \frac{m}{M} - \frac{m}{2M} = \frac{m}{2M}$.

Therefore, it eventually holds:

$$\left|\frac{1}{\hat{I}(t)}\right| = \left|\frac{1}{I(t)} + \frac{I(t) - \hat{I}(t)}{I(t)\hat{I}(t)}\right| \leq \frac{M}{m} + \frac{2M^2}{m^2}B_6(n-1)^{-2}$$

So, for large enough n, we obtain $A_I = \frac{|D^1\hat{I}(t) - D^1 I(t)|}{\hat{I}(t)} \leq \left(1 + \frac{M}{m}\right)B_5(n-1)^{-1}$.

As for A_{II}, we have $A_{II} = \frac{|D^1 I(t)| \cdot |I(t) - \hat{I}(t)|}{I(t)\hat{I}(t)}$. As shown in the proof of Proposition 2, it holds $\left|D^1 I(t)\right| \leq \frac{2M^2}{m^2}$ for all large enough n, and by applying the bounds obtained for A_I above the following inequality is readily obtained:

$$A_{II} = \frac{|D^1 I(t)| \cdot |I(t) - \hat{I}(t)|}{I(t)\hat{I}(t)} \leq \frac{\frac{2M^2}{m^2} \times M}{m}\left(1 + \frac{M}{m}\right)B_6(n-1)^{-2}$$

Therefore, there exists $B_7 < \infty$ such that, for each t in $[0, 1]$, $\left|\frac{D^1\hat{I}(t)}{\hat{I}(t)} - \frac{D^1 I(t)}{I(t)}\right| \leq B_7(n-1)^{-1}$.

By the same procedure as in Proposition 2 it is shown that, for each t in $[0, 1]$, it holds $I(t) > 0$, and the same is true for $\hat{I}(t)$ for large enough n. This ensures that

$$\left|\hat{LTE}_n - LTE\right| = \left|\int_0^1 \left(D^1 \ln \hat{I}(t) - D^1 \ln I(t)\right) dt\right| \leq \int_0^1 \left|D^1 \ln \hat{I}(t) - D^1 \ln I(t)\right| dt$$

$$\leq B_7(n-1)^{-1} \int_0^1 dt = B_5(n-1)^{-1}$$

for sufficiently large n, which completes the proof of both parts of the proposition.

Proof of Proposition 4

First, we will prove (a), that for each n large enough and each (fixed) set of knots $N_n = \{t_i = (i-1)/(n-1), i = 1, \ldots, n\}$, the vector of interpolated time paths, $\hat{Z}_n = (\hat{z}_{1,n}, \ldots, \hat{z}_{2r,n})$, is $\mathbf{B}(S)$-measurable. We shall use symbol $(.)_n$ to denote the operator that associates with each path in $C^1[0,1]$ its natural spline interpolant, with n knots located at N_n. Thus, $\hat{z}_{j,n} = (z_j)_n$ is the natural spline that interpolates path z_j at N_n.

It suffices to show that $(.)_n$ is a continuous mapping of $C^1[0,1]$ into $C^1[0,1]$, which implies that it is also continuous as a vector function of S into S, and therefore $\hat{Z}_n = (Z)_n$ is $\mathbf{B}(S)$-measurable. Select two arbitrary functions $f, f' \in C^1[0,1]$, with that function space endowed with the norm $\|.\|$ particularized to the case of a single trajectory. We will show that, given n fixed (and therefore, a fixed set of n knots), $\|f - f'\| \to 0$ implies $\|\hat{f} - \hat{f}'\| \to 0$. This stems, as we shall see, from the fact that $(.)_n$ is a continuous linear operator.

Indeed it can be shown that, provided that $n \geq 2$, the set of natural splines (having degree 3, continuous derivatives up to order 1 and knots at N_n) is a vector space of dimension n. This implies that the natural spline interpolant $\hat{f}_n(t)$ for $f \in C^1[0,1]$ is unique (Powell 1981, Chap. 23, Theorem 23.1.) and has the following expression:

$$\hat{f}_n(t) = \sum_{i=1}^n \hat{\beta}_{i,n} \varphi_{i,n}(t)$$

where $(\varphi_{1,n}, \ldots, \varphi_{n,n})$ is a vector of n linearly independent functions in $C^1[0,1]$. The coefficient vector $\hat{\beta}_n = \left(\hat{\beta}_{1,n}, \ldots, \hat{\beta}_{n,n}\right)^T$ is obtained by imposing the n conditions of interpolation at N_n, i.e., $f(t_i) = \hat{f}(t_i)$, $i = 1, \ldots, n$. This is equivalent to solving the system of linear equations $f_n = \Phi_n \hat{\beta}_n$, where $f_n = (f(t_1), \ldots, f(t_n))^T$ and $\Phi_n = [c_{i,k}]$, $i, k = 1, \ldots, n$, is a squared matrix whose elements are $c_{i,k} = \varphi_{i,n}(t_k)$. Uniqueness of the solution of this problem implies that matrix Φ_n is nonsingular, so $\hat{\beta}_n = \Phi_n^{-1} f_n$.

Now consider another function $f' \in C^1[0,1]$, and let $\hat{f}'_n(t) = \sum_{i=1}^{n} \hat{\beta}'_{i,n}\varphi_{i,n}(t)$ be its natural spline interpolant at N_n, with $\hat{\beta}'_n = \left(\hat{\beta}'_{1,n}, \ldots, \hat{\beta}'_{n,n}\right)^T$. For any integer $0 \leq \alpha \leq 1$, select an arbitrary point $0 \leq t \leq 1$. We have

$$\left|D^\alpha f'_n(t) - D^\alpha \hat{f}_n(t)\right| = \left|\sum_{i=1}^{n} D^\alpha \varphi_{i,n}(t)\left(\hat{\beta}'_{i,n} - \hat{\beta}_{i,n}\right)\right| \leq \sum_{i=1}^{n}\left|D^\alpha \varphi_{i,n}(t)\right| \cdot \left|\hat{\beta}'_{i,n} - \hat{\beta}_{i,n}\right| \leq$$

$$\max_{0 \leq \alpha \leq 1} \max_{i=1,\ldots,n}\left|D^\alpha \varphi_{i,n}(t_i)\right| \cdot \sum_{i=1}^{n}\left|\hat{\beta}'_{i,n} - \hat{\beta}_{i,n}\right| \leq \max_{0 \leq \alpha \leq 1} \max_{i=1,\ldots,n} \max_{t \in [0,1]}\left|D^\alpha \varphi_{i,n}(t)\right| n\, n^{-1} \sum_{i=1}^{n}\left|\hat{\beta}'_{i,n} - \hat{\beta}_{i,n}\right| \leq$$

$$\max_{0 \leq \alpha \leq 1} \max_{i=1,\ldots,n} \max_{t \in [0,1]}\left|D^\alpha \varphi_{i,n}(t)\right| n\sqrt{n^{-1} \sum_{i=1}^{n}\left(\hat{\beta}'_{i,n} - \hat{\beta}_{i,n}\right)^2}$$

As $\qquad \hat{\beta}'_n - \hat{\beta}_n = \Phi_n^{-1} e_n,$ $\qquad$ with $\qquad e_n = f'_n - f_n,$ $\qquad$ we $\qquad$ have
$$\sqrt{\left(\hat{\beta}'_n - \hat{\beta}_n\right)^T\left(\hat{\beta}'_n - \hat{\beta}_n\right)} = \sqrt{e_n^T\left(\Phi_n^{-1}\right)^T \Phi_n^{-1} e_n}.$$

Since matrix $\left(\Phi_n^{-1}\right)^T \Phi_n^{-1}$ is symmetric and positive definite, it admits the orthogonal decomposition $\left(\Phi_n^{-1}\right)^T \Phi_n^{-1} = P_n^T \Lambda_n P_n$, where $P_n^T P_n = I_n$, with I_n being the unit matrix of order n and Λ_n being a diagonal matrix with the eigenvalues of $\left(\Phi_n^{-1}\right)^T \Phi_n^{-1}$ on its main diagonal. It is readily obtained that $\left[e_n^T\left(\Phi_n^{-1}\right)^T \Phi_n^{-1} e_n\right]^{1/2} = \left[e_n^T P_n^T \Lambda_n P_n e_n\right]^{1/2} \leq \lambda_1^{1/2}\left[e_n^T e_n\right]^{1/2}$, where λ_1 is the largest eigenvalue of $\left(\Phi_n^{-1}\right)^T \Phi_n^{-1}$.

Therefore, we obtain

$$\sqrt{\left(\hat{\beta}_n - \hat{\beta}_n\right)^T\left(\hat{\beta}_n - \hat{\beta}_n\right)} = \sqrt{e_n^T\left(\phi_n^{-1}\right)^T \phi_n^{-1} e_n} \leq \sqrt{\lambda_1}\sqrt{e_n^T e_n} = \sqrt{\lambda_1}\sqrt{\sum_{i=1}^{n}\left(f'(t_i) - f(t_i)\right)^2}$$

$$\leq \sqrt{\lambda_1 n} \max_{i=1,\ldots,n}\left|f'(t_i) - f(t_i)\right| \leq \sqrt{\lambda_1 n} \cdot \|f' - f\|$$

and it holds $\left|D^\alpha \hat{f}'_n(t) - D^\alpha \hat{f}_n(t)\right| \leq \max_{0 \leq \alpha \leq 1} \max_{i=1,\ldots,n} \max_{t \in [0,1]}\left|D^\alpha \varphi_{i,n}(t)\right| n\sqrt{\lambda_1} \|f' - f\|.$ Since both α and t are arbitrary the above bound is uniform, so $\|\hat{f}'_n - \hat{f}_n\| \to 0$ as $\|f' - f\| \to 0$, which ensures continuity—in terms of the distance induced by the norm $\|f\| = \max_{0 \leq \alpha \leq 1} \max_{0 \leq t \leq 1}|D^\alpha f(t)|$ in $C^1[0,1]$—of the natural spline interpolant $\hat{f}_n$. This evidently implies that the mapping from S into S defined by the natural spline interpolation operator with knots at N_n, applied element by element of $Z = (z_1, \ldots, z_{2r})$, i.e., $\hat{Z}_n = (Z)_n = \left((z_1)_n, \ldots, (z_{2r})_n\right)$, is continuous with respect to the metric induced by the norm $\|Z\| = \max_{j=1,\ldots,2r} \max_{0 \leq \alpha \leq 1} \max_{0 \leq t \leq 1}|D^\alpha z_j(t)|$. (For brevity we use the same symbol, $\|.\|$, for the norms of

$C^1[0, 1]$ and S; the proper interpretation will be clear in each case depending on the context.)

Since we have assumed that Z—the vector of time paths—is $\mathbf{B}(S)$-measurable, the approximant vector $\hat{Z}_n$, generated by natural spline interpolation, is also $\mathbf{B}(S)$-measurable, as it is obtained by a continuous (and so, measurable) transformation of Z.

Part (b) of the statement is obtained directly. For $n \geq 2$ the natural spline interpolant $\hat{Z}_n$ is unique and $\mathbf{B}(S)$-measurable, as established in part (a). Select an arbitrary point $\omega \in \Omega$. By Assumption $1'$.(i) each component of the observed path vector $Z(., \omega)$ belongs to $W^2[0,1]$, and Lemma A.1 establishes, given $\omega \in \Omega$, uniform convergence with respect to t, i.e., $\left\| \hat{Z}_n(., \omega) - Z(., \omega) \right\| \to 0$ as $n \to \infty$. Since Assumption $1'$.(ii) imposes that, for some $m > 0$ and each $\omega \in \Omega$, it holds $\min_{0 \leq t \leq 1} D^1 z_j(t, \omega) \geq m$, it is then obtained that, for each j and all large enough n (possibly depending on ω), it holds (uniformly in t) $D^1 \hat{z}_j(t, \omega) = D^1 z_j(t, \omega) + \left(D^1 \hat{z}_j(t, \omega) - D^1 z_j(t, \omega) \right) \geq m - m/2 = m' > 0$. This is a consequence of $\min_{j=1,...,2r} \min_{0 \leq t \leq 1} D^1 z_j(t, \omega) \geq m$, which is ensured by Assumption $1'$.(ii) and the fact that $\max_{0 \leq t \leq 1} \left| D^1 \hat{z}_j(t, \omega) - D^1 z_j(t, \omega) \right| \to 0$ as $n \to \infty$ by Lemma A.1. Therefore, for each $\omega \in \Omega$ and n large enough, it holds $\min_{j=1,...,2r} \min_{0 \leq t \leq 1} D^1 \hat{z}_j(t, \omega) \geq m' > 0$, that is, $\hat{Z}_n(., \omega) \in A_{m'}$.

It is easily checked that, for any $m' > 0$, the mappings $g(Z)$ and $h(Z)$ defining, respectively, TE and LTE, are continuous (and thus $\mathbf{B}(S)$-measurable) in $A_{m'}$, which in turn is a closed subset of S with nonempty interior, made up of all the vector functions Z in S with coordinates belonging to $W^2[0, 1]$ and having $\min_{j=1,...,2r} \min_{0 \leq t \leq 1} D^1 \hat{z}_j(t, \omega) \geq m'$ for some $m' > 0$ fixed a priori and not depending on ω.

So, for each $\omega \in \Omega$ and n large enough, the sample realization of the interpolant for Z also has first derivative that is (uniformly in $[0,1]$) bigger than some $m' > 0$, i.e., $\hat{Z}_n(\omega) \in A_{m'}$. Therefore, with probability 1, it holds $\hat{Z}_n \in A_{m'}$ as $n \to \infty$, and $\hat{T}E_n = g(\hat{Z}_n)$ and $\hat{L}TE_n = h(\hat{Z}_n)$ are continuous (and so $\mathbf{B}(\Re)$-measurable) functions of $\hat{Z}_n$ for each n large enough.

Once we have established that, for large enough n, $\hat{T}E_n$ and $\hat{L}TE_n$ are random variables, convergence with probability 1 to TE y LTE, respectively, is derived by the same procedure as in Propositions 2 y 3, applied to an arbitrary realization $\omega \in \Omega$. For each $\omega \in \Omega$, the integrals appearing in the definitions of TE and LTE are classical Riemann integrals, so the proofs of Lemmas A.1–A.3 and Propositions 2 and 3 readily extend (for each fixed $\omega \in \Omega$) to the random case, with the only inconsequential issue that the Lipschitz bounds $(B_1, \ldots, B_6)$ will generally range with each realization ω of the random experiment.

References

Adams RA (1975) Sobolev spaces. Academic Press, New York

Billingsley P (1968) Convergence of probability measures. Wiley, New York

DeVore RA, Lorentz GG (1993) Constructive approximation. Springer, Berlin

Dudley RM (1973) Sample functions of the Gaussian process. Annals of Probability 1(1):66–103

Fernández Vázquez E, Fernández González P (2008) An extension to Sun's decomposition methodology: the path based approach. Energy Economics 30(3):1020–1036

Powell MJD (1981) Approximation theory and methods. Cambridge University Press, Cambridge

Schoenberg I (1964) Spline functions and the problem of graduation. Proceedings of the National Academy of Sciences USA 52:947–950

Schultz M (1973) Spline Analysis. Prentice Hall, Engelwood Cliffs

Tanaka K (1996) Time series analysis: nonstationary and noninvertible distribution theory. Wiley, New York

Wahba G (1990) Spline models for observational data. SIAM, Philadelphia

Chapter 3
Multiplicative Decomposition of the Change in Aggregate Energy Intensity in the European Countries During the 1995–2010 Period

3.1 Introduction

In this chapter a multiplicative decomposition of the variation of aggregate energy intensity in the European economy (EU27) will be conducted. An analysis of the specific factors influencing those variations is also included. As in other methodological (Boyd et al. 1988; Howarth et al. 1991) and empirical studies (Jenne and Cattell 1983; Li et al. 1990; Gardner 1993), we will focus on the *energy intensity approach*.

First we shall outline two general parametric Divisia-index-based methods proposed by Liu et al. (1992), including six specific cases developed and refined by Ang (1995) and Ang and Choi (1997). Then we will apply the described methodology, along with the splines method studied in Chap. 2, in order to decompose the change in aggregate energy intensity at a sectoral disaggregation level. A brief analysis of energy prices is also included with the aim of facilitating the interpretation of some results.

3.2 Methodology

3.2.1 Energy Intensity Approach. The Multiplicative Case

Total energy intensity is defined as the ratio of total energy consumption in industry to total industrial production. Following the energy intensity approach (Ang 1995), the variations in that ratio can be explained in terms of the contributions from two factors: (a) technological change and variations in the use of high quality energy inputs (so-called *intensity effect*), and (b) changes in the structure of production (*structural effect*).

Following Ang and Lee (1994), we shall base our analysis on the following set of variables, evaluated at time t:

P. Fernández González et al., *The Driving Forces of Change in Environmental Indicators*, Lecture Notes in Energy 25, DOI: 10.1007/978-3-319-07506-8_3, © Springer International Publishing Switzerland 2014

e_t: Total energy consumption.
$e_{j,t}$: Energy consumption in economic sector j.
y_t: GDP (in ppp) in the EU27.
$y_{j,t}$: Production in sector j.
$S_{j,t}$: Production share in sector j (calculated as $S_{j,t}=y_{j,t}/y_t$).
I_t: Aggregate energy intensity ($I_t=e_t/y_t$).
$I_{j,t}$: Energy intensity in sector j ($I_{j,t}=e_{j,t}/y_{j,t}$).

In terms of sectorally disaggregated data, the following decomposition is readily obtained:

$$e_t = \sum_{j=1}^{r} e_{j,t} = \sum_{j=1}^{r} y_t \left(y_{j,t}/y_t\right) \left(e_{j,t}/y_{j,t}\right) = \sum_{j=1}^{r} y_t S_{j,t} I_{j,t} \qquad (3.1)$$

where r is the number of economic sectors considered in the sectoral disaggregation.

3.2.1.1 Standard Divisia-Index-Based Methods

By considering infinitesimal periods, dividing by I_t and integrating on both sides of the equation with respect to time on interval $[0,T]$, the following decomposition is obtained for the percentage change in energy intensity:

$$\ln(I_T/I_0) = \int_0^T \sum_{j=1}^{r} \frac{S'_{j,t} I_{j,t}}{I_t}\, dt + \int_0^T \sum_{j=1}^{r} \frac{S_{j,t} I'_{j,t}}{I_t}\, dt \qquad (3.2)$$

where $S'_{j,t}$ and $I'_{j,t}$ are, respectively, the first derivatives of $S_{j,t}$ and $I_{j,t}$ with respect to time.

Denoting by R_{tot} the total effect ($R_{tot} = I_T/I_0$), and by R_{str} and R_{int} the structural and intensity effects, Eq. (3.2) can be expressed in the following two alternative forms:

$$R_{tot} = \exp\left(\int_0^T \sum_{j=1}^{r} M_{j,t}\left(S'_{j,t}/S_{j,t}\right) dt\right) \exp\left(\int_0^T \sum_{j=1}^{r} M_{j,t}\left(I'_{j,t}/I_{j,t}\right) dt\right) = R_{str} R_{int}$$

$$(3.3)$$

where $M_{j,t} = e_{j,t}/e_t$, and

$$R_{tot} = \exp\left(\int_0^T \sum_{j=1}^{r} \left(I_{j,t} S'_{j,t}/I_t\right) dt\right) \exp\left(\int_0^T \sum_{j=1}^{r} \left(I'_{j,t} S_{j,t}/I_t\right) dt\right) = R_{str}\, R_{int} \quad (3.4)$$

Assuming that the above path integrals satisfy certain conditions (Liu et al. 1992), it is possible to transform the line integral into a parametric problem with only a finite number of free parameters (namely, β_j and τ_j; $j = 1, \ldots, r$), to be detailed below. Liu et al. (1992) propose the following general parametric Divisia methods:

(a) Parametric Divisia Method 1 (PDM1):

$$R_{str} = \exp\left(\sum_{j=1}^{r} \left[M_{j,0} + \beta_j (M_{j,T} - M_{j,0}) \right] \ln\left(\frac{S_{j,T}}{S_{j,0}} \right) \right) \qquad (3.5)$$

$$R_{int} = \exp\left(\sum_{j=1}^{r} \left[M_{j,0} + \tau_j (M_{j,T} - M_{j,0}) \right] \ln\left(\frac{I_{j,T}}{I_{j,0}} \right) \right) \qquad (3.6)$$

(b) Parametric Divisia Method 2 (PDM2):

$$R_{str} = \exp\left(\sum_{j=1}^{r} \left[\left(\frac{I_{j,0}}{I_0} \right) + \beta_j \left(\frac{I_{j,T}}{I_T} - \frac{I_{j,0}}{I_0} \right) \right] (S_{j,T} - S_{j,0}) \right) \qquad (3.7)$$

$$R_{int} = \exp\left(\sum_{j=1}^{r} \left[\left(\frac{S_{j,0}}{I_0} \right) + \tau_j \left(\frac{S_{j,T}}{I_T} - \frac{S_{j,0}}{I_0} \right) \right] (I_{j,T} - I_{j,0}) \right) \qquad (3.8)$$

with $0 \leq \beta_j \leq 1$ and $0 \leq \tau_j \leq 1$.

Due to practical impossibility to obtain observations in continuous time, the product of Eqs. (3.5)–(3.6) or (3.7)–(3.8) will generally differ from the exact[1] value of R_{tot}, so inclusion of a residual component, R_{rsd}, is required. The following final expression is readily obtained for the decomposition of the change in aggregate energy intensity:

$$R_{tot} = R_{str} R_{int} R_{rsd} \qquad (3.9)$$

The parameter values (β_j and τ_j) in the above expressions are, respectively, the weights of the corresponding variables at periods 0 and T. Values for those coefficients can be assigned in a number of ways, resulting in different specific decomposition methods. In this chapter some of those particular approaches will be applied to multiplicatively decompose general structures of both the PDM1 and PDM2 types. More specifically, we will implement the following parametric methods: LAS-PDM1,[2] AVE-PDM1,[3] LAS-PDM2, AVE-PDM2, and AWT-PDM.

---

[1] The exception is the additive AVE-PDM2 method, which produces exact decompositions.

[2] As mentioned in Chap. 1, this method is regarded as "Laspeyres-based" because values from the initial period are used as weights in Eqs. (3.5)–(3.8).

[3] The term "simple average" indicates that the same weight (0.5) is assigned to the two extreme periods 0 and T.

We will also consider a nonparametric Divisia method, the *refined Divisia method* (or LMDI), outlined in Chap. 1. As commented above, this is a zero-residual method having many nice mathematical properties. Relying on an energy demand approach, Ang and Choi (1997) proposed the following weight function:

$$L\left(M_{j,0}, M_{j,T}\right) = \frac{\left(M_{j,T} - M_{j,0}\right)}{\ln\left(M_{j,T}/M_{j,0}\right)} \tag{3.10}$$

As seen in Chap. 1, the following normalization is introduced in order to ensure that the sum of weights is equal to unity,

$$w_j^* = \frac{L\left(M_{j,0}, M_{j,T}\right)}{\sum_{j=1}^{r} L\left(M_{j,0}, M_{j,T}\right)} \tag{3.11}$$

where w_j^* denotes[4] the normalized weight of economic sector $j = 1,..., r$.

3.2.1.2 The Splines Method

This is another nonparametric technique, also providing exact decompositions. As outlined in Chap. 2, cubic splines are applied to reconstruct/interpolate the paths of sectoral energy consumptions $(e_{j,t})$ and productions $(y_{j,t})$, and the analogy principle is then applied to obtain reconstructions for the trajectories of total production and sector shares/intensities. Finally, numerical integration is used to compute[5] the approximate cumulative effects, as detailed in the previous chapter.

3.2.2 Time Series Decomposition

As seen in Chap. 1, this methodology allows researchers to perform decompositions in which information available for intermediate periods is exploited. In the multiplicative decomposition case, if the cumulative change in overall energy intensity (*total effect*) in the $[0,T]$ period is denoted by $(C_{\text{tot}})_{0,T}$, the cumulative structural effect by $(C_{\text{str}})_{0,T}$, the cumulative intensity effect by $(C_{\text{int}})_{0,T}$, and the

---

[4] The Excel file 'Chapter3.xls' in the accompanying extra contents includes implementations of all the above methods.

[5] Matlab codes for the splines method as applied in this chapter are included in the m-file 'Chapter3splines.m' in the accompanying extra contents.

cumulative residual term is denoted by $(C_{rsd})_{0,T}$, then the cumulative effects for the period may be factored in terms of interannual changes in the following way:

$$(C_{tot})_{0,T} = (R_{tot})_{0,1} (R_{tot})_{1,2} \cdots (R_{tot})_{T-1,T} \tag{3.12}$$

$$(C_{str})_{0,T} = (R_{str})_{0,1} (R_{str})_{1,2} \cdots (R_{str})_{T-1,T} \tag{3.13}$$

$$(C_{int})_{0,T} = (R_{int})_{0,1} (R_{int})_{1,2} \cdots (R_{int})_{T-1,T} \tag{3.14}$$

$$(C_{rsd})_{0,T} = (R_{rsd})_{0,1} (R_{rsd})_{1,2} \cdots (R_{rsd})_{T-1,T} \tag{3.15}$$

3.3 Analysis of the Change in Aggregate Energy Intensity in the European Union

Now we will carry out a decomposition of the change in energy intensity in the European economy during the 1995–2010 period. Time series of sectoral energy consumption and production in the EU-27 were obtained from Eurostat (2013). Specifically, the following sectors will be considered:

- *Sector 1*. Agriculture and fisheries.
- *Sector 2*. Industry and Construction.
- *Sector 3*. Services (excepting transport).
- *Sector 4*. Transport.

3.3.1 Graphical Analysis

Before applying the methodology outlined in Sect. 3.2, a graphical analysis is useful in order to obtain an idea of the time patterns displayed by energy consumption, output and energy intensity in the 1995–2010 period.

First, a significant reduction in total energy intensity is observed throughout the period (see Fig. 3.1), particularly in Sectors 4 (Transport) and 2 (Industry and Construction). However, the evolution of energy intensity in Sector 3 (Services) looks striking as this is the only sector that did not reduce energy intensity and, in addition, its behaviour was relatively invariable and constant along the study period. The remaining sectors maintained a downward trend, although spikes are observed in some years such as 2003 (Sectors 1, 2 and 3), 2009 (Sector 4) and 2010 (Sectors 1 and 2), possibly related to the increase in oil prices and situations of economic and financial crisis.

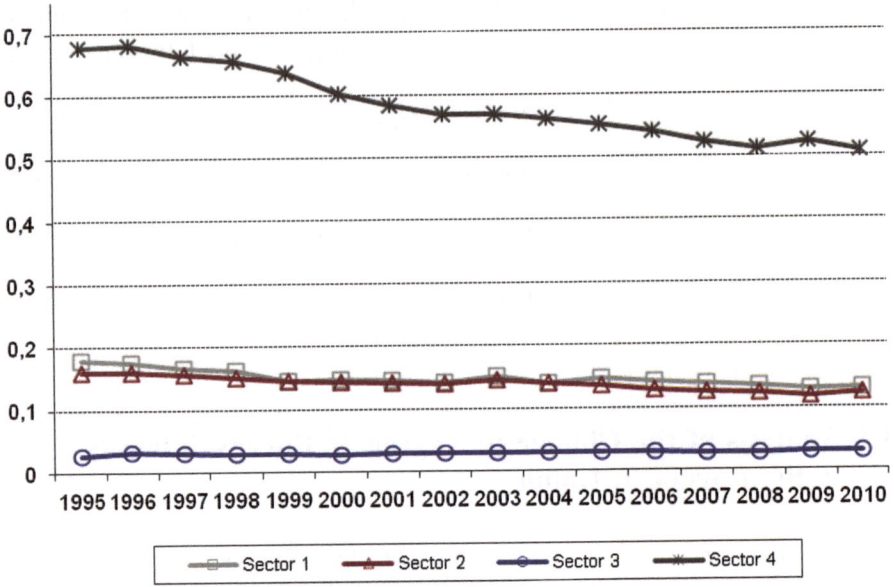

Fig. 3.1 Energy intensity in the EU27 (thousand toe per million of euro) by economic sector (1995–2010)

As for the evolution of production (see Fig. 3.2) and energy intensity (Fig. 3.1), Sector 2 (Industry and Construction) reduced its energy intensity, and slightly diminished its production quota. On the contrary, Sector 4 (Transport) reduced energy intensity at the same time its production quota increased. These remarks anticipate a great relevance of the intensity effect. Furthermore, Sector 3 (Services) slightly increases both its intensity and its production quota, which would highlight the relevance of structural change to the variation in the aggregate. Summing up, preliminary graphical analysis suggests the influence of both effects, intensity and structural, on the reduction in aggregate energy intensity that was observed in the European economy during the 1995–2010 period. However, since Sectors 2 and 4 were the largest consumers of energy (see Fig. 3.3), their respective weights will be larger, so the intensity effect may be expected to prevail over the structural effect. In any event, a more rigorous decomposition analysis is required before drawing any conclusions.

3.3.2 Contributions of the Determinant Factors

Then the decomposition techniques outlined above were applied in order to perform a multiplicative decomposition of the aggregate.

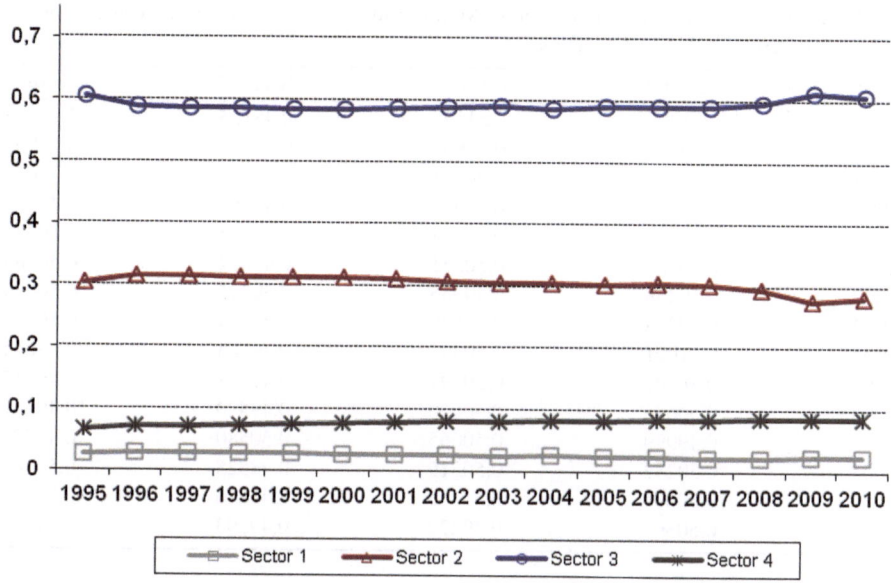

Fig. 3.2 Participation of economic sectors in the European production (1995–2010)

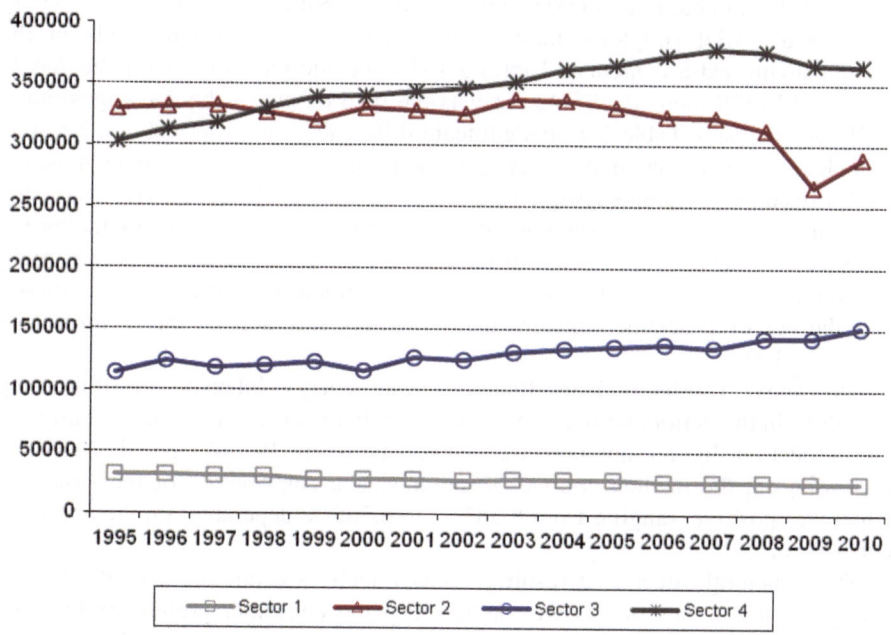

Fig. 3.3 Power consumption in the EU27 (in thousand toe) by economic sector (1995-2010)

Table 3.1 β_j values by economic sector. Multiplicative time series decomposition for the 1996–2010 period. AWT-PDM method

Year	Sector 1	Sector 2	Sector 3	Sector 4
1996	0.51390	0.51173	0.48135	0.50895
1997	0.50836	0.49930	0.51077	0.49965
1998	0.50288	0.50601	0.49779	0.49692
1999	0.52147	0.50565	0.49463	0.49931
2000	0.49259	0.49392	0.51775	0.50373
2001	0.50385	0.50351	0.48034	0.50489
2002	0.50488	0.49955	0.50396	0.50099
2003	0.49368	0.49678	0.49584	0.50423
2004	0.51001	0.50497	0.49765	0.49854
2005	0.48793	0.50294	0.49738	0.49917
2006	0.50685	0.50620	0.49590	0.49611
2007	0.49999	0.50065	0.50580	0.49874
2008	0.50282	0.50319	0.48525	0.50183
2009	0.50004	0.51115	0.48846	0.48978
2010	0.50368	0.49222	0.49293	0.51121

3.3.2.1 Divisia-Index-Based Decomposition Methods

In order to enable a comparison of results and detect any potential differences among techniques, the following battery of specific decomposition methods was successively applied: LAS-PDM1, AVE-PDM1, LAS-PDM2, AVE-PDM2, AWT-PDM, and LMDI (implementation details appear in file Chapter3.xls in the accompanying extra contents). Tables 3.1–3.3 include the weights for the AWT-PDM and LMDI methods. Table 3.4 lists the final output of the decompositions.

When analysing Table 3.4 no significant differences are observed in the results provided by the various methods. As expected, Laspeyres-based methods have the largest residual deviations (Ang and Lee 1994), whereas LMDI delivers an exact decomposition. The results obtained through PDM1-based methods, including the adaptive weights one, are similar to those of the LMDI approach. This fact might be explained by the fact that the relationship between the variable to decompose and the set of predefined factors approximately has a logarithmic form (see Figs. 3.1–3.3).

The above results indicate that aggregate energy intensity has fallen by 13.278 % in the period, with the intensity effect being negative and contributing to that reduction by an amount that ranges between 19.744 % and 20.510 % depending on the method applied. In contrast, the contribution of the structural effect was positive, ranging from 8.387 % to 8.765 % depending on the approach considered.

The structural effect is a result of issues such as economic growth and the changes in trade patterns, as well as the growing consumer preference for high added value products and services that incorporate improved and low-energy-intensity materials. Another factor that may have an indirect influence on the

Table 3.2 τ_j values by economic sector. Multiplicative time series decomposition for the 1996–2010 period. AWT-PDM method

Year	Sector 1	Sector 2	Sector 3	Sector 4
1996	0.49991	0.50648	0.51023	0.49965
1997	0.49854	0.49589	0.50088	0.49128
1998	0.50369	0.50017	0.49647	0.49168
1999	0.50075	0.49915	0.49528	0.48930
2000	0.50107	0.49181	0.50100	0.49066
2001	0.50626	0.50275	0.49189	0.49645
2002	0.50115	0.49982	0.49840	0.49266
200₃	0.51315	0.50322	0.50025	0.50341
2004	0.48995	0.49892	0.49821	0.49409
2005	0.50887	0.49921	0.49516	0.49486
2006	0.50568	0.49621	0.49293	0.49185
2007	0.50337	0.49505	0.49610	0.49109
2008	0.49313	0.50453	0.49123	0.49666
2009	0.48179	0.51565	0.48694	0.49387
2010	0.50918	0.49531	0.50218	0.50488

Table 3.3 Normalized weights (w_j^*) by economic sector. Multiplicative time series decomposition for the 1996–2010 period. LMDI method

Year	Sector 1	Sector 2	Sector 3	Sector 4
1996	0.03965	0.41898	0.15090	0.39048
1997	0.03882	0.41479	0.15168	0.39471
1998	0.03769	0.41012	0.14894	0.40325
1999	0.03577	0.39969	0.15111	0.41342
2000	0.03445	0.39997	0.14719	0.41839
2001	0.03394	0.40105	0.14782	0.41718
2002	0.03298	0.39583	0.15281	0.41838
2003	0.03237	0.39649	0.15328	0.41787
2004	0.03207	0.39456	0.15528	0.41810
2005	0.03198	0.38798	0.15678	0.42326
2006	0.03113	0.38041	0.15917	0.42929
2007	0.02988	0.37507	0.15868	0.43636
2008	0.02959	0.36907	0.16214	0.43920
2009	0.03039	0.34825	0.17314	0.44821
2010	0.03061	0.34051	0.18081	0.44806

structural effect is energy price, as it can affect the competitive positions of the various economic sectors depending on their higher or lower energy intensities.

On the contrary, the intensity effect relates to technological change, innovation, capital-labour-energy substitution, and use of higher quality energy. Again, energy price is a factor to consider, as it may either stimulate or inhibit investment in new technologies, best adaptation to available techniques, the search for more energy efficient alternatives, as well as R&D to enable higher margins and the best prices

Table 3.4 Multiplicative time series decomposition of the change in aggregate energy intensity in the EU27, from 1995 to 2010. Base year 1995

Method	C_{tot}	C_{str}	C_{int}	C_{rsd}
LAS-PDM1	0.86722	1.08387	0.79490	1.00656
AVE-PDM1	0.86722	1.08600	0.79854	1.00001
LAS-PDM2	0.86722	1.08765	0.80256	0.99350
AVE-PDM2	0.86722	1.08604	0.79853	0.99998
AWT-PDM	0.86722	1.08602	0.79856	0.99997
LMDI	0.86722	1.08601	0.79854	1.00000

Table 3.5 Results of multiplicative LMDI decomposition of the variation in energy intensity in the European economy (1995–2010). Interannual effects

Year	R_{str}	R_{int}	R_{tot}
1995	1	1	1
1996	1.03756	1.02502	1.06352
1997	1.00923	0.97003	0.97899
1998	1.00604	0.97821	0.98413
1999	1.01163	0.96785	0.97910
2000	1.00895	0.95886	0.96744
2001	1.00403	0.99183	0.99583
2002	1.00602	0.98010	0.98600
2003	0.99789	1.01918	1.01703
2004	1.00786	0.97923	0.98692
2005	0.99995	0.98241	0.98236
2006	1.00150	0.96528	0.96673
2007	1.00368	0.96312	0.96666
2008	0.99801	0.99150	0.98953
2009	0.98397	0.99934	0.98333
2010	1.00725	1.00763	1.01494

available (Fernández González and Pérez Suárez 2000). Results in Table 3.4 clearly show, regardless of the method applied, the high relevance of the intensity effect in reducing aggregate energy intensity.

Tables 3.5 and 3.6 report, respectively, the interannual and cumulative results provided by the LMDI method.[6]

The graphical representation of the estimated cumulative effects in Fig. 3.4 below facilitates a study of their evolution. A growing trend in the impact of both factors, especially noticeable in the intensity effect, is observed throughout the analysed period. The overall result, a significant reduction in aggregate energy intensity, suggests a greater contribution of the intensity effect. However, a more detailed analysis reveals some years when changes are observed in the behaviour of the effect. In particular, in years 1996, 2003 and 2010, aggregate energy intensity grew up, mainly as a consequence of the intensity effect.

---

[6] Our choice of LMDI method was motivated by its zero-residual character.

Table 3.6 Results of multiplicative LMDI decomposition of the variation in energy intensity in the European economy (1995–2010). Cumulative effects

Year	C_{str}	C_{int}	C_{tot}
1995	1.00000	1.00000	1.00000
1996	1.03756	1.02502	1.06352
1997	1.04714	0.99430	1.04117
1998	1.05347	0.97264	1.02465
1999	1.06572	0.94137	1.00324
2000	1.07526	0.90264	0.97058
2001	1.07960	0.89527	0.96653
2002	1.08610	0.87745	0.95300
2003	1.08380	0.89428	0.96923
2004	1.09232	0.87571	0.95655
2005	1.09227	0.86030	0.93968
2006	1.09391	0.83043	0.90842
2007	1.09793	0.79981	0.87813
2008	1.09575	0.79301	0.86894
2009	1.07819	0.79249	0.85446
2010	1.08601	0.79854	0.86722

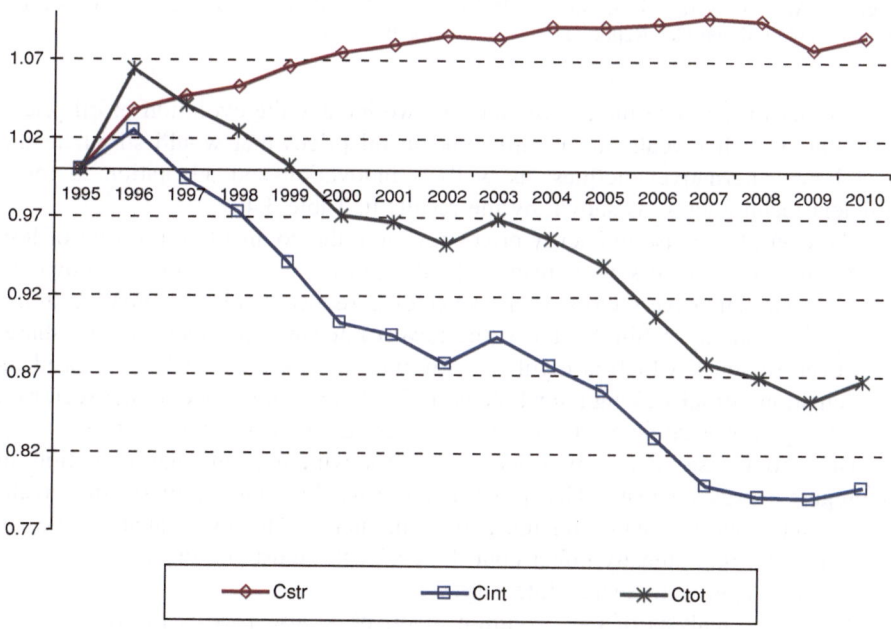

Fig. 3.4 Multiplicative LMDI decomposition (cumulative effects) of the variation in aggregate energy intensity in the European economy (1995–2010). Base year 1995

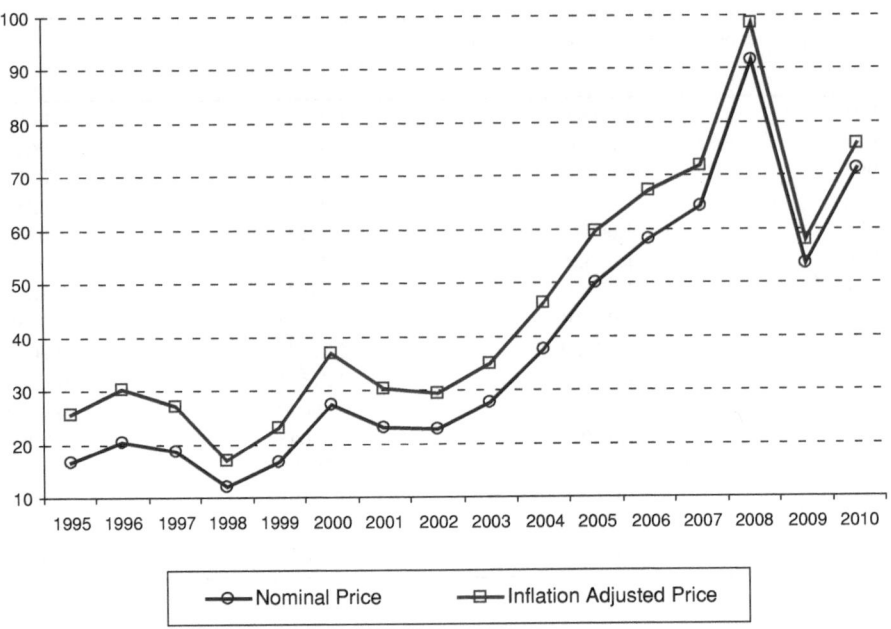

Fig. 3.5 Annual Average Domestic Crude Oil Prices (in $/Barrel), years 1995–2010. *Source:* Saudi Arabia Oil and Gas Report, Business Monitor International

In order to facilitate the above analysis, we include the evolution of oil prices (see Fig. 3.5). It reveals an upward trend in oil prices that would stimulate the search for alternative energies, as well as innovation and adaptation to more efficient technologies, as shown by the results in Table 3.4.

The overall increase in energy prices favoured the competitive position of less energy-intensive sectors with respect to the most intensive ones, as shown in Fig. 3.2. The intensive sectors, such as Sector 2, reduced their participation in the European economy, while Sector 3 (the lowest intensity one) increased its share. However Sector 4, which is highly energy-intensive (see Fig. 3.1), increased its participation virtually along the whole period. The great influence of that sector on the aggregate was sufficient to offset the lowering effects of other sectors.

The strong rise in energy prices coincides with a significant reduction in aggregate energy intensity. That price increase would stimulate innovation, technical change and the search for more efficient energy. The above analysis clearly indicates that the intensity effect contributed, in the most significant way, to the reduction in aggregate energy intensity.

A detailed analysis of the evolution of oil price also reveals the presence of inflection points at certain times. Indeed, a rebound in oil prices is observed in the first years of study, as a result of strong growth in the US and Asia economies that lasted until 1997, when the Asian financial crisis brings about a price decrease. In year 2003, a new surge occurs, mainly due to the Second Gulf War; oil price did

Table 3.7 Results of multiplicative time series decomposition of the change in aggregate energy intensity in the European economy between 1995 and 2010. Base year 1995

Method	C_{tot}	C_{str}	C_{int}	C_{rsd}
Splines	0.86722	1.08594	0.79859	1

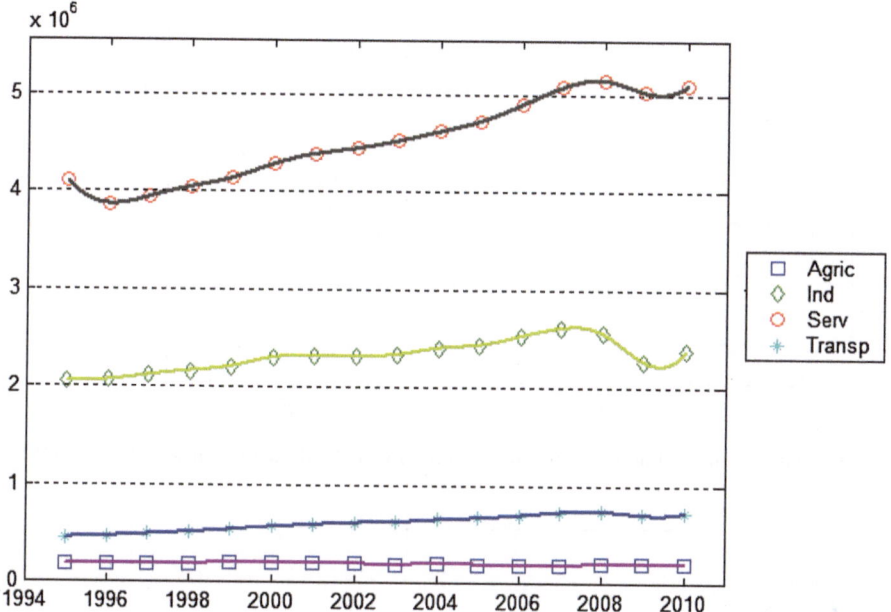

Fig. 3.6 Time paths (spline-based reconstructions) for sectoral production (millions of euro)

not stop growing until the financial crisis of 2008, sharply dropping in 2009. Finally, in 2010, the economies reviving after the Great Recession and the events of the Arab Spring lead to further price increases. This analysis would lead us to relate the decreases in oil price with smaller negative contributions of the intensity effect and a reduction in the positive influence of the structural effect. It is also reasonable to assume that lower oil prices could reduce the production costs of enterprises and alleviate their need to invest in technologies that are more efficient from an energy standpoint.

3.3.2.2 Spline-Based Decomposition

The effects estimated when the method of splines is applied to the above decomposition are detailed in Table 3.7. Figures 3.6–3.10 include the reconstructed trajectories for sectoral energy consumptions (e_j) and productions (y_j), as well as the cumulative effects (intensity, structural, total).

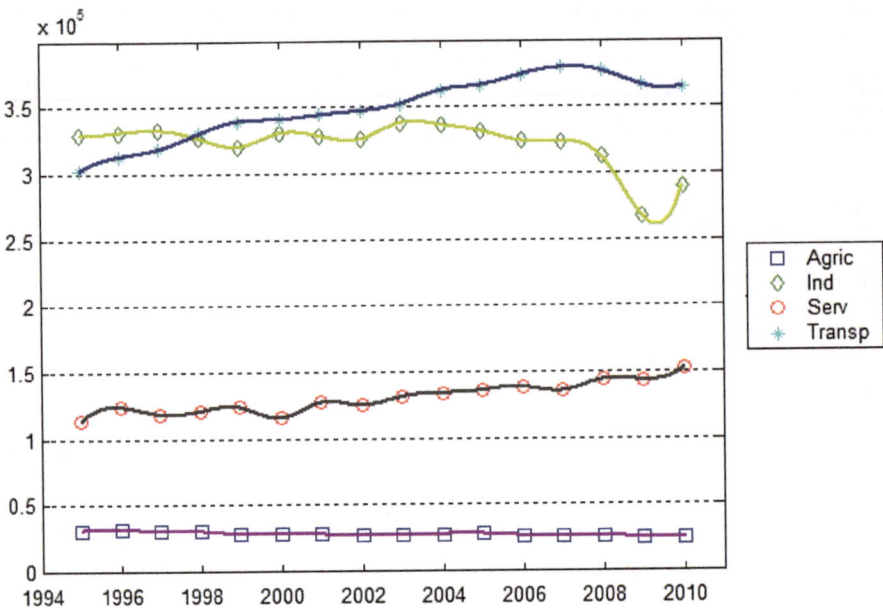

Fig. 3.7 Time paths (spline-based reconstructions) for sectoral energy consumption in MWh

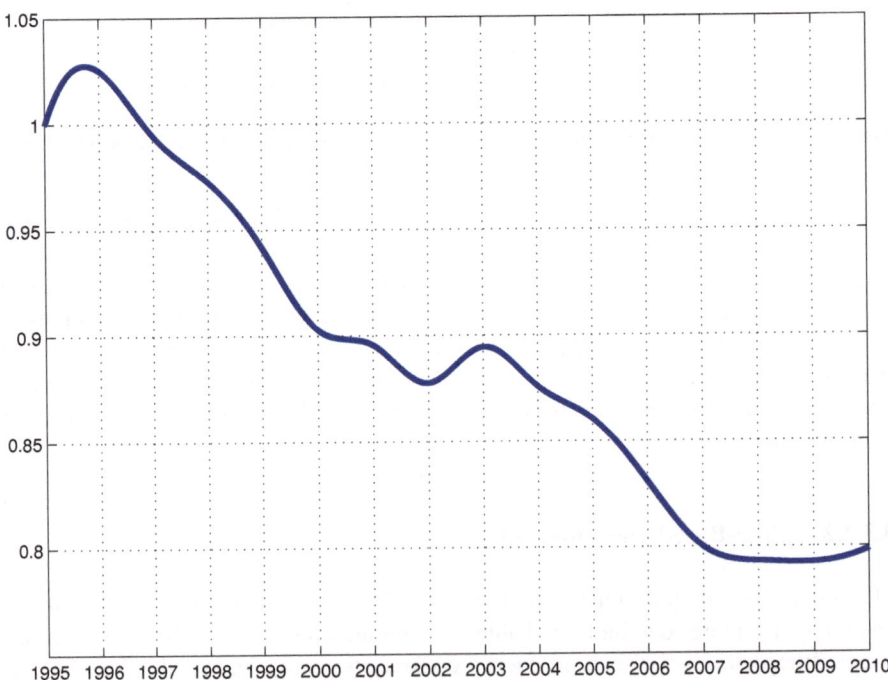

Fig. 3.8 Evolution of the cumulative intensity effect (C_{int}) for the multiplicative (spline-based) decomposition of the variation in aggregate energy intensity

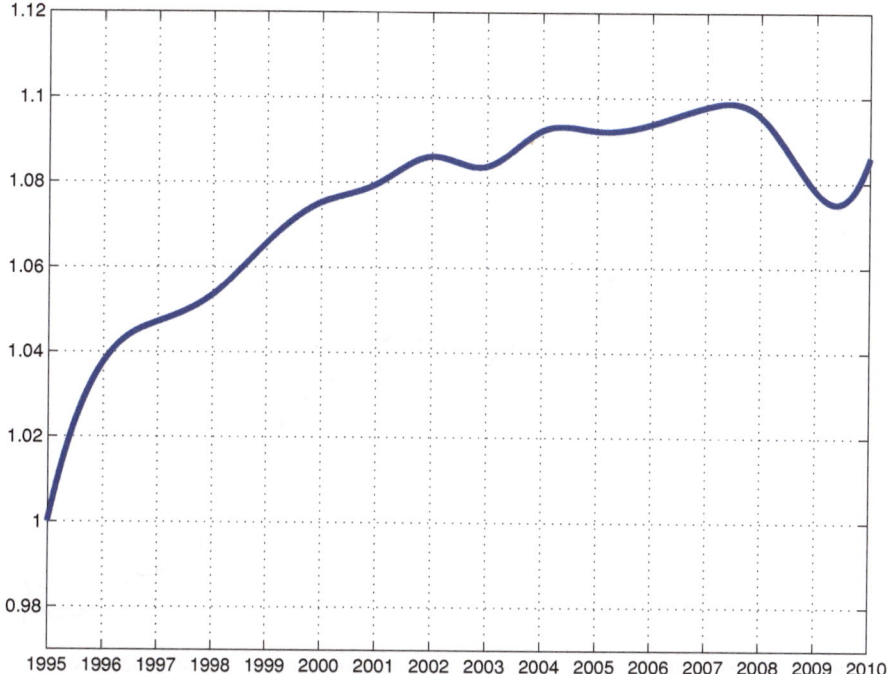

Fig. 3.9 Evolution of the cumulative structural effect (C_{str}) for the multiplicative (spline-based) decomposition of the variation in aggregate energy intensity

The production trajectories in Fig. 3.6 show different patterns depending on sector and the study period. During the first half of the period production increased in all sectors. However, in the second half some sectors (e.g., Group 1, Agriculture and Fisheries) exhibit a decrease in their growth rates, and the effects of the Great Recession after 2008 are also particularly visible in Groups 2 (Industry) and 3 (Services). In any event, all the groups experienced production increases in the 1995–2010 period.

With regard to energy consumption (Fig. 3.7), a gradual decrease is observed in Sector 1, whereas in Group 2 (Industry and Construction) a sharp drop is observed, particularly after 2008 as a consequence of the recession that so strongly affected that group. On the contrary, in Sector 3 (Services), and particularly in Sector 4 (Transport) there is an increase in energy consumption between years 1995 and 2010, although the effects of the last recession are also clearly visible in the Transport sector.

Figures 3.8–3.10 show, respectively, the time evolution of the intensity, structural and total effects for the multiplicative decomposition of the change in aggregate energy intensity in the European economy. These cumulative effects are obtained as detailed in Chap. 2, i.e., by integration between periods 0 and t, $0 \leq t \leq 1$, where $t = 0$ represents year 1995 and $t = 1$ corresponds to 2010.

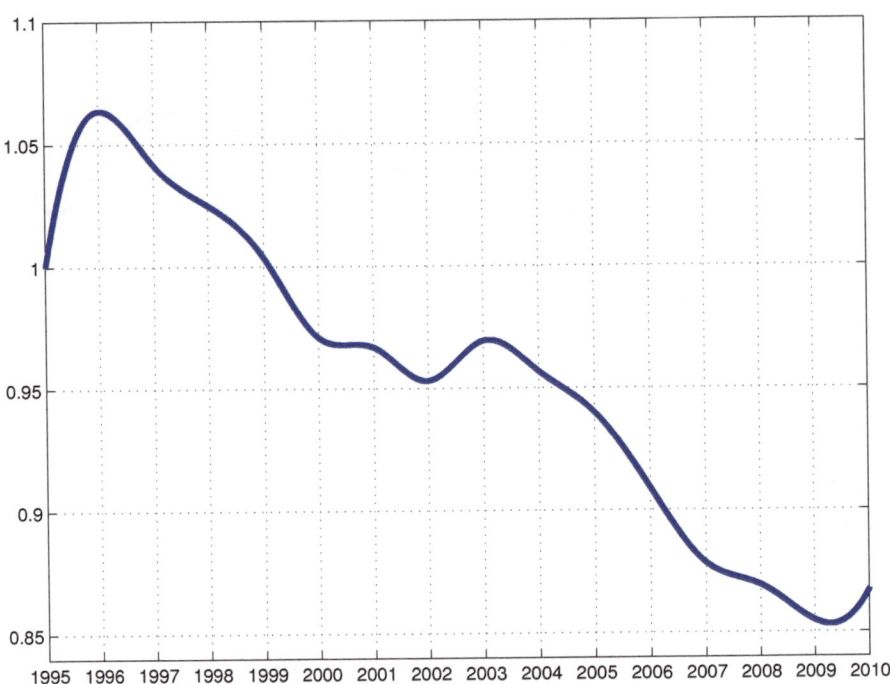

Fig. 3.10 Evolution of the cumulative total effect (C_{tot}) for the multiplicative (spline-based) decomposition of the variation in aggregate energy intensity

As expected, in the overall analysis for the 1995–2010 period a negative intensity effect is observed, which helped reducing aggregate energy intensity in the European economy. On the contrary, the structural effect was positive, thus tending to increase the aggregate. Again, the influence of the intensity effect was larger than that of the structural effect, resulting in a negative overall effect. Figures 3.8–3.10 show the evolution of the various cumulative effects, again pointing out the relevance of innovation, technical change, adaptation to new technologies and the use of higher quality energy, in order to reduce aggregate energy intensity in Europe. Not surprisingly, a clear similarity between Figs. 3.8–3.10 and Fig. 3.4 (obtained by applying the LMDI method discussed above) is also observed.

Figures 3.8–3.10 also suggest the presence of several phases in the evolution of the aggregate. Thus, in the first half of the period (excluding year 1996), aggregate energy intensity experienced a sharp decline, due solely to the intensity effect. In that period, the structural effect was positive and increasing, tending to dampen the drop induced by the intensity effect. In the second half of the period (with the exception of year 2003), aggregate energy intensity drops again as a consequence of the intensity effect. In this case, although being positive, the structural effect reduces its importance in the aggregate.

The cumulative effects estimated by the splines method are reported in Table 3.7 below. When comparing Tables 3.5 and 3.7 a remarkable similarity is observed between the results provided by the splines method and those of conventional techniques (in particular the LMDI method). From that perspective, standard decompositions can be seen as simplified forms of the more sophisticated approach proposed in Chap. 2. We may conjecture that use of longer time series would result in a greater disparity in the results provided by the various methods, although more extensive analysis is required to properly test that hypothesis.

3.4 Conclusions

Aggregate energy intensity experienced a significant reduction in the European economy along the study period. Whichever decomposition method is applied, that drop appears to be almost exclusively a result of the intensity effect, as the structural effect only contributed significantly to that reduction in year 2009. The structural effect accumulated along the whole period was positive, thus helping increase the aggregate.

Energy prices have been included in our analysis, as they play an important role in improving energy efficiency and can help us understand the decomposition results. No doubt the increase in energy prices during the study period had a direct effect on the structure of production in most sectors, decreasing production in energy-intensive sectors (excluding transport). On the other hand, the energy price increases have probably encouraged investment in search for more efficient techniques and higher quality energy.

The splines method was also applied in this chapter. This is a nonparametric approach, and when its results are compared with those provided by LMDI (which is also nonparametric in nature), it is observed that both techniques lead to exact decompositions with very similar results.

The above results indicate that factors such as technical change, innovation, diffusion and adaptability to more efficient technologies have been key drivers for the reduction in aggregate energy intensity in the EU27. On the contrary, the same evidence suggests that we cannot rely on the change in the production structure as a reducing force for the aggregate.

References

Ang BW (1995) Decomposition methodology in industrial energy demand analysis. Energy 20(11):1081–1095

Ang BW, Choi KH (1997) Decomposition of aggregate energy and gas emission intensities for industry: a refined Divisia index method. Energy J 18(3):59–73

Ang BW, Lee SY (1994) Decomposition of industrial energy consumption: some methodological and application issues. Energy Econ 16(2):83–92

Boyd G, Hanson DA, Sterner T (1988) Decomposition of changes in energy intensity: a comparison of the Divisia index and other methods. Energy Econ 10(4):309–312

Fernández González P, Pérez Suárez R (2000) Descomposición de la intensidad de energía agregada industrial en España: el efecto estructural y el efecto intensidad. XIV Reunión ASEPELT, Oviedo (Spain)

Gardner DT (1993) Industrial energy use in Ontario from 1962 to 1984. Energy Econ 15(1):25–32

Howarth RB, Schipper L, Duerr PA, Strøm S (1991) Manufacturing energy use in eight OECD countries. Energy Econ 13(2):135–142

Jenne C, Cattell R (1983) Structural change and energy efficiency in industry. Energy Econ 5(2):114–123

Li JW, Shrestha RM, Foell WK (1990) Structural change and energy use: the case of the manufacturing sector in Taiwan. Energy Econ 12(2):109–115

Liu XQ, Ang BW, Ong HL (1992) The application of the Divisia index to the decomposition of changes in industrial energy consumption. Energy J 13(4):161–177

Chapter 4
Additive Decomposition of Changes in Greenhouse Gas Emissions in the European Union in the 1990s

4.1 Introduction

After classical decomposition techniques, including those based on Laspeyres, Paasche and Marshall-Edgeworth weighting schemes, new models[1] have been developed. Ang (2004) compares several decomposition methods and includes some implementation recommendations.

In this chapter we incorporate two additive techniques which provide exhaustive alternatives to LMDI and the splines method. These are respectively the solution proposed by Sun (1998), focused on the 'jointly created and equally distributed' principle, and the path-based (PB) approach derived by Fernández Vázquez and Fernández González (2008). The latter provides an interesting alternative in those cases where there is additional information available in intermediate periods for some of the variables. That information is used within a maximum entropy (ME) estimation procedure in order to estimate the parameters that determine the time paths of the decomposition factors.

Once the above methodologies are briefly outlined, they will be applied to an empirical analysis of changes in greenhouse gas emissions in the EU15 from 1990 to 2002, considering as determinants of that change the GDP of the EU15, the share of each country in the European GDP and the intensity of greenhouse gas emissions. We will obtain exhaustive additive decompositions through splines, Sun's, PB and LMDI methods, under three alternative scenarios. In particular, three initial information situations will be assumed, depending on the availability of: (a) only the values of the variables in the extreme periods, (b) triennial series of all variables (or at least some of them in the PB approach) or (c) annual series of all variables. The results for cases (b) and (c), where the information available is richer, will be compared with those obtained for (a), with the aim of checking which of the methods applied, and under what circumstances, provides a 'more accurate' solution.

---

[1] For instance, Albrecht et al. (2002) introduced a technique based on the Kaya identity that leads to exhaustive, symmetric decompositions.

P. Fernández González et al., *The Driving Forces of Change in Environmental Indicators*, Lecture Notes in Energy 25, DOI: 10.1007/978-3-319-07506-8_4, © Springer International Publishing Switzerland 2014

4.2 Methodology

We shall consider an endogenous variable V, whose value is given by the product of n exogenous variables or determinants, namely:

$$V = x_1 x_2 \ldots x_n \tag{4.1}$$

We assume that each determinant can change without necessarily affecting the values of the other determinants, and that the increases in V are time dependent.

Focusing on an additive decomposition, the variation of V between times 0 and 1 is readily calculated as

$$\Delta V = V^1 - V^0 = x_1^1 x_2^1 \ldots x_n^1 - x_1^0 x_2^0 \ldots x_n^0 \tag{4.2}$$

where x_i^t is the value of x_i at time t; $i = 1, \ldots, n$, $t = 0, 1$.

Considering the particular case in which there are only two factors, x and y, we have:

$$\Delta V = V^1 - V^0 = x^1 y^1 - x^0 y^0 \tag{4.3}$$

The following expressions suggest two possible ways to decompose (4.3), namely:

$$\Delta V = y^0 \Delta x + x^0 \Delta y + \Delta x \Delta y \tag{4.4}$$

and

$$\Delta V = y^1 \Delta x + x^1 \Delta y - \Delta x \Delta y \tag{4.5}$$

Equation (4.4) uses a Laspeyres weighting structure, while (4.5) would correspond to a Paasche decomposition. Both expressions are approximate, non-exhaustive decompositions, since the sum of the main effects of the factors not necessarily equals the change in the endogenous variable V. There generally appears an interaction effect, $\Delta x \Delta y$, which can be seen as a residual term. In principle exact decompositions are preferable to non-exhaustive ones, except in cases where only a few factors are analysed and interpretation of the residual terms is unambiguous and has a practical interest. As seen above, the LMDI method is an exhaustive decomposition technique. However, from a theoretical standpoint numerous, approximately equivalent formulas exist whose implementation usually leads to results that differ to greater or lesser extent. Given non-uniqueness in the results, in this chapter we will discuss two exhaustive alternatives to LMDI. More specifically, Sun's method and the PB approach rely on the idea of a priori specifying a functional form for the trajectories of the factors.

4.2.1 Sun's Method

As a general rule, specifying a certain time path for the determinants amounts to selecting a specific way to distribute the interaction terms in the decomposition.

In this regard, Sun's method (1998)[2] applies the principle of 'jointly created and equally distributed'. That principle suggests that the part of total variation that is due to the interaction of two or more effects must be equally distributed between them. Thus, in the two-factor case, the researcher considers the average of the two extreme paths (*PP*1 and *PP*2), which implies an equal division of the interaction rectangle (see Fig. 4.1). Application of this principle leads to complete decompositions without residual term, in which the interaction of the factors is assigned to the main effects.

In the general case (4.1) where V is the product of n determinants, the effect of changes in each factor is given by the following expression (Sun 1998):

$$
\begin{aligned}
D_i = \Delta x_i \text{Effect} = &\int_0^1 \prod_{j \neq i}^n x_j \frac{dx_i}{dt} dt = \left[\prod_{j<i}^{i-1} x_j^0 \right] \Delta x_i \left[\prod_{j<i}^n x_j^0 \right] \\
&+ \sum_{j \neq i}^n \left[\frac{1}{2} \prod_{k<i}^{i-1} x_k^0 \Delta x_i \prod_{i<k<j}^{j-1} x_k^0 \Delta x_j \prod_{k>j}^n x_k^0 \right] \\
&+ \sum_{j \neq i}^n \sum_{l \neq j,i}^n \left[\frac{1}{3} \prod_{k<i}^{i-1} x_k^0 \Delta x_i \prod_{i<k<j}^{j-1} x_k^0 \Delta x_j \prod_{j<k<l}^{l-1} x_k^0 \Delta x_l \prod_{k>l}^n x_k^0 \right] \\
&\cdots \\
&+ \frac{1}{n} \left[\prod_{j=1}^n \Delta x_j \right]
\end{aligned}
\tag{4.6}
$$

By construction, it holds $\sum_{i=1}^n \Delta x_i$ Effect $= \Delta V$, so this is an exhaustive method that eliminates the residual term by splitting it among the set of determinants x_i.

Our starting point in this chapter is the equation that expresses the amount of greenhouse gas emissions in country j (denoted by c_j) as the product of three factors: the GDP of the whole EU15 (y), the share of the GDP of country j in the GDP of the whole EU15 (S_j) and, finally, the greenhouse gas emissions of country j per unit of national GDP (I_j). The relationship may be expressed in the following form:

$$
c_j = y \, S_j I_j
\tag{4.7}
$$

where $S_j = \frac{y_j}{\sum_{j=1}^{15} y_j} = \frac{y_j}{y}$ and $I_j = \frac{c_j}{y_j}$, with y_j being the GDP of country j.

---

[2] Within input–output analysis, Dietzenbacher and Los (1998) have proposed a similar solution for structural decomposition analysis.

Fig. 4.1 Polar (*PP*1 and
*PP*2) versus linear paths (*LP*)

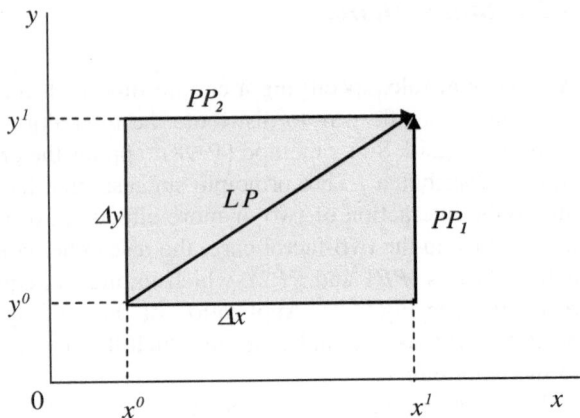

Our goal is to additively decompose the change (D_{tot}) in emissions in each
country between years 1990 and 2002 into the following three components:

$$D_{tot} = D_{act} + D_{str} + D_{int} \tag{4.8}$$

where D_{act} (the activity effect) is the contribution of changes in y to the variation of
c_j, and D_{int} (the intensity effect) and D_{str} (the structural effect) denote, respectively,
the contributions of changes in S_j and I_j. According to Eq. (4.6) above, these
contributions can be expressed, respectively, as:

$$D_{act} = I_j^{1990} S_j^{1990} \Delta y + \frac{1}{2} \Delta I_j S_j^{1990} \Delta y + \frac{1}{2} I_j^{1990} \Delta S_j \Delta y + \frac{1}{3} \Delta I_j \Delta S_j \Delta y \tag{4.9}$$

$$D_{str} = I_j^{1990} \Delta S_j y^{1990} + \frac{1}{2} \Delta I_j \Delta S_j y^{1990} + \frac{1}{2} I_j^{1990} \Delta S_j \Delta y + \frac{1}{3} \Delta I_j \Delta S_j \Delta y \tag{4.10}$$

$$D_{int} = \Delta I_j S_j^{1990} y^{1990} + \frac{1}{2} \Delta I_j \Delta S_j y^{1990} + \frac{1}{2} \Delta I_j S_j^{1990} \Delta y + \frac{1}{3} \Delta I_j \Delta S_j \Delta y \tag{4.11}$$

where $\Delta y = y^{2002} - y^{1990}$, $\Delta I_j = I_j^{2002} - I_j^{1990}$, and $\Delta S_j = S_j^{2002} - S_j^{1990}$.

4.2.2 The Path-Based Method

Within the field of SDA, Fernández Vázquez (2004) proposed an exhaustive
decomposition technique that he denominates path-based (PB) method. That
technique is connected with the work of Harrison et al. (2000), who develop ideas
of Vogt (1978). Later, in Fernández Vázquez and Fernández González (2008), the
method is extended to the IDA field. Among other advantages (including its

relative flexibility), it allows use of additional information (even though it may be partial, or it is only available for a single factor), thereby reducing the degree of arbitrariness of the analysis.

The PB approach assumes that the determinants (x_i) change continuously over time (t), which varies between 0 and 1. Therefore, we have

$$V(t) = x_1(t)x_2(t)...x_n(t) \tag{4.12}$$

Assuming differentiability of each factor $(x_i(t))$, an infinitesimal change in V can be expressed as:

$$dV = \frac{\partial V}{\partial x_1}\frac{dx_1}{dt}dt + \cdots + \frac{\partial V}{\partial x_n}\frac{dx_n}{dt}dt \tag{4.13}$$

Finally, the total change in V can be expressed as a 'sum' of infinitesimal changes between 0 and 1, or more rigorously

$$\Delta V = \int_0^1 \frac{dV}{dt}dt = \int_0^1 \sum_{i=1}^n \frac{\partial V}{\partial x_i}\frac{dx_i}{dt}dt \tag{4.14}$$

The effects of the determinants are:

$$\Delta x_i \text{ Effect} = \int_0^1 \frac{\partial V}{\partial x_i}\frac{dx_i}{dt}dt = \int_0^1 \prod_{j \neq i}^n x_j \frac{dx_i}{dt}dt, \ i = 1,\ldots,n \tag{4.15}$$

Equation (4.15) shows that the derivatives of the factors with respect to t play a central role in determining the magnitude of the effects attributed to the changes in those determinants. Accordingly, the specification of the time path for each factor $(x_i(t) = f_i(t))$ between the initial and final periods may have a large impact on the measurement of the effects. Harrison et al. (2000) proposed assuming linear paths for all the variables, which amounts to the solution derived from Sun's method (1998). However, the values of the factors at intermediate points along the paths may differ from those of the straight line. The PB method relaxes the linearity assumption and allows for more flexible functional forms that make explicit use of available information, in order to distribute the interaction effects among the main effects of the determinants. In particular, the PB approach considers the following class of trajectories:

$$x_i(t) = x_i^0 + \Delta x_i t^{\theta_i}; 0 \leq t \leq 1, \ \theta_i > 0 \tag{4.16}$$

The above class contains monotonic functions going from x^0 to x^1 and having no inflection points (see Fig. 4.2).

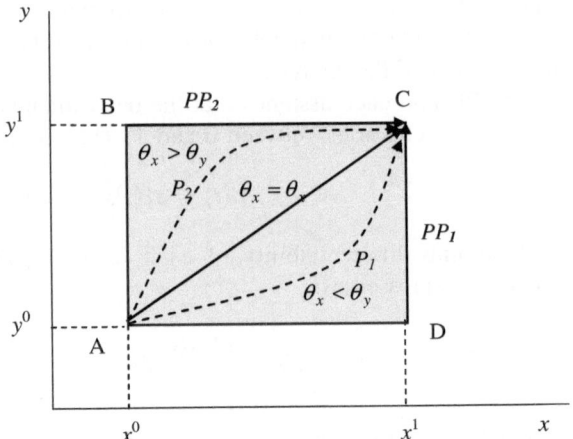

Fig. 4.2 Some generalized monotonic time paths allowed by the PB approach

The main idea in the PB approach is shaped by parameter θ_i, which also determines the fraction of the interaction effect that is attributed to each determinant. Paths $PP1$ (if $\theta_x/\theta_y \to 0$), $PP2$ (if $\theta_x/\theta_y \to \infty$), and the linear trajectory connecting points A and C, are extreme cases in this class of functions, with paths $P1$ and $P2$ being examples of some intermediate situations.

By replacing the time paths (4.16) into Eq. (4.15), the following expression is obtained:

$$
\Delta x_i \text{ Effect} = \int_0^1 \prod_{j \neq i}^n x_j \frac{dx_i}{dt} \, dt = \left[\prod_{j<i}^{i-1} x_j^0 \right] \Delta x_i \left[\prod_{j>i}^n x_j^0 \right]
$$

$$
+ \sum_{j \neq i}^n \left[\frac{\theta_i}{\theta_i + \theta_j} \prod_{k<i}^{i-1} x_k^0 \Delta x_i \prod_{i<k<j}^{j-1} x_k^0 \Delta x_j \prod_{k>j}^n x_k^0 \right]
$$

$$
+ \sum_{j \neq i}^n \sum_{l \neq j, i}^n \left[\frac{\theta_i}{\theta_i + \theta_j + \theta_l} \prod_{k<i}^{i-1} x_k^0 \Delta x_i \prod_{i<k<j}^{j-1} x_k^0 \Delta x_j \prod_{j<k<l}^{l-1} x_k^0 \Delta x_l \prod_{k>l}^n x_k^0 \right]
$$

$$
\cdots
$$

$$
+ \frac{\theta_i}{\sum_{j=1}^n \theta_j} \left[\prod_{j=1}^n \Delta x_j \right]
$$

$$
(4.17)
$$

The higher the value of θ_i is in comparison with the remaining θ_j values, the greater is the share of the interaction effect that is attributed to x_i, and therefore its contribution to the change in aggregate V will be larger.

Certainly, when there is no information available on the time evolution of the determinants, a plausible hypothesis would be that the parameters of all the paths are equal($i.e.$, $\theta_1 = \theta_2 = \cdots = \theta_n$). However, if information on the evolution of all factors is available for some intermediate periods, a dynamic time series decomposition would enable a more efficient use of the data. Unfortunately, most times we only have partial information available for some determinant factors. For those cases, Fernández Vázquez and Fernández González (2008) proposed a methodology that exploits partial information, generally leading to more accurate results than when that information is ignored. Their idea relies on estimating the parameters of the time paths by the method of generalized maximum entropy (GME). The objective is to fit the parameters of the following linear model,[3] where the dependent variable (y) is 'explained' in terms of a set of H covariates:

$$y = X\theta + e \qquad (4.18)$$

where y is a ($T \times 1$) vector of observations, X is a ($T \times H$) matrix of explanatory variables for each observation, θ is the ($H \times 1$) vector of parameters to be fitted, and e is a ($T \times 1$) vector of random disturbances. Model (4.18) can be fitted by the method proposed by Golan et al. (1996).

In our case the PB method assumes the following trajectories for the factors:

$$I_j(t) = I_j^{1990} + \Delta I_j t^{\theta_{I_j}}, \qquad (4.19)$$

$$S_j(t) = S_j^{1990} + \Delta S_j t^{\theta_{S_j}}, \qquad (4.20)$$

where $j = 1, \ldots, 15$, and

$$y(t) = y^{1990} + \Delta y t^{\theta_y} \qquad (4.21)$$

Then the θ_is are estimated, and by particularizing expression (4.17) the following expressions are readily obtained for the effects:

$$D_{\text{act}} = I_j^{1990} S_j^{1990} \Delta y + \frac{\theta_y}{\theta_{I_j} + \theta_y} \Delta I_j S_j^{1990} \Delta y + \frac{\theta_y}{\theta_{S_j} + \theta_y} I_j^{1990} \Delta S_j \Delta y$$
$$+ \frac{\theta_y}{\theta_{I_j} + \theta_{S_j} + \theta_y} \Delta I_j \Delta S_j \Delta y \qquad (4.22)$$

---

[3] A detailed description of the model and its estimation appears in Fernández Vázquez and Fernández González (2008).

$$D_{str} = I_j^{1990} \Delta S_j \, y^{1990} + \frac{\theta_{S_j}}{\theta_{I_j} + \theta_{S_j}} \Delta I_j \, \Delta S_j \, y^{1990} + \frac{\theta_{S_j}}{\theta_{S_j} + \theta_y} I_j^{1990} \Delta S_j \, \Delta y$$

$$+ \frac{\theta_{S_j}}{\theta_{I_j} + \theta_{S_j} + \theta_y} \Delta I_j \, \Delta S_j \, \Delta y \qquad\qquad (4.23)$$

$$D_{int} = \Delta I_j \, S_j^{1990} y^{1990} + \frac{\theta_{I_j}}{\theta_{I_j} + \theta_{S_j}} \Delta I_j \, \Delta S_j \, y^{1990} + \frac{\theta_{I_i}}{\theta_{I_j} + \theta_y} \Delta I_j \, S_j^{1990} \Delta y$$

$$+ \frac{\theta_{I_j}}{\theta_{I_j} + \theta_{S_j} + \theta_y} \Delta I_j \, \Delta S_j \, \Delta y \qquad\qquad (4.24)$$

It is observed that when the components of θ are equal for all factors (i.e. $\theta_{I_j} = \theta_{S_j} = \theta_y$, $\forall j$) the solution proposed by Sun (1998) is obtained.

4.2.3 The LMDI Method

By following the energy consumption approach introduced by Ang and Lee (1994), and relying on the LMDI method, we can readily derive an additive decomposition to study the variations in aggregate greenhouse gas emissions. As commented above, we shall focus on a decomposition that includes three components summarizing, respectively, the effects of (a) aggregate output (*activity effect*, D_{act}), (b) inter-regional structural changes (*structural effect*, D_{str}) and (c) the proportion of quality energy inputs and technical change (*intensity effect*, D_{int}).

In terms of disaggregated regional data, the following decomposition is obtained:

$$c_t = \sum_{j=1}^{r} c_{j,t} = \sum_{j=1}^{r} y_t \left(y_{j,t}/y_t \right) \left(c_{j,t}/y_{j,t} \right) = \sum_{j=1}^{r} y_t S_{j,t} I_{j,t} \qquad (4.25)$$

where r is the number of regions (respectively countries).

Focusing on the total effect, $D_{tot} = c_t - c_0$ (the variation of greenhouse gas emissions between 0 and t) and relying on Eq. (4.25), we can decompose D_{tot} as follows:

$$c_t - c_0 = \sum_{j=1}^{r} w_j(t) \ln\left(\frac{y_t}{y_0}\right) + \sum_{j=1}^{r} w_j(t) \ln\left(\frac{S_{j,t}}{S_{j,0}}\right) + \sum_{j=1}^{r} w_j(t) \ln\left(\frac{I_{j,t}}{I_{j,0}}\right) \qquad (4.26)$$

where $w_j(t)$ is a weighting function.

If we consider the proposal by Ang and Choi (1997), the above weight function is given by the following expression:

$$w_j(t) = L(c_{j,0}, c_{j,t}) = \frac{c_{j,t} - c_{j,0}}{\ln(c_{j,t}) - \ln(c_{j,0})} \tag{4.27}$$

Therefore, we have[4]

$$D_{\text{act}} = \sum_{j=1}^{r} \left(\frac{c_{j,t} - c_{j,0}}{\ln(c_{j,t}) - \ln(c_{j,0})} \right) \ln (y_t/y_0) \tag{4.28}$$

$$D_{\text{str}} = \sum_{j=1}^{r} \left(\frac{c_{j,t} - c_{j,0}}{\ln(c_{j,t}) - \ln(c_{j,0})} \right) \ln (S_{j,t}/S_{j,0}) \tag{4.29}$$

$$D_{\text{int}} = \sum_{j=1}^{r} \left(\frac{c_{j,t} - c_{j,0}}{\ln(c_{j,t}) - \ln(c_{j,0})} \right) \ln (I_{j,t}/I_{j,0}) \tag{4.30}$$

$$D_{\text{tot}} = D_{\text{act}} + D_{\text{str}} + D_{\text{int}} \tag{4.31}$$

Implementation details for LMDI and Sun's methods appear in file Chapter4.xls in the accompanying extra materials.

4.2.4 The Splines Method

As detailed in Chap. 2, this method relies on a reconstruction of the trajectories of productions and energy consumptions by country. Approximate time paths are generated by cubic spline interpolation, and reconstructions for the trajectories of total production and country shares/intensities are then obtained as in Chap. 3. Numerical integration[5] is then used to compute plug-in approximants for the cumulative effects, as seen in Chap. 2.

4.3 Analysis of Changes in Greenhouse Gas Emissions in the EU15 in the 1990s

In this section we will study the contributions of the three factors mentioned above (respectively, GDP of the EU15, the weight of each country in the European GDP and intensity of greenhouse gas emissions) to the change in greenhouse gas

---

[4] This kind of decomposition provides a significant refinement over conventional (parametric) Divisia methods.

[5] Matlab codes for the splines method as applied in this chapter are included in the m-file 'Chapter4splines.m' in the accompanying extra contents.

emissions in the EU15[6] between years 1990 and 2002. The data were obtained from Eurostat (New Cronos data base) and coincide with those previously analysed by Fernández Vázquez and Fernández González (2008) and Fernández González (2013).

4.3.1 Contributions of Changes in the Factors

We begin with the additive decomposition of the variation in total emissions. In order to compare results we will apply the various approaches (Sun's, PB, LMDI, splines), considering three potential scenarios, each representing a different case in terms of available information, namely:

(i) *Annual time series decomposition*: yearly information is available. This will be the 'most favourable' case in terms of information.

(ii) *Triennial decomposition*: only triennial information is available. This is an intermediate case in terms of data availability.

(iii) *Periodwise decomposition*: the researcher only has information for the initial and final years (respectively 1990 and 2002) of the study period. This evidently is the worst case scenario.

We opted for triennial information in order to maintain the same structure as Fernández Vázquez and Fernández González (2008). This facilitates comparison with results from LMDI and splines methods, also allowing us to study an intermediate situation between the 'least informative' (periodwise decomposition) and the 'most informative' (annual time series decomposition) scenarios.

In addition, for the implementation of the PB approach we will consider that additional, triennial information is only available for one of the variables. In particular, we assume that the national GDP for each country in the EU15 is only known for years 1993, 1996 and 1999. Therefore, in the intermediate years, information is only available for factors S_j and y. On the contrary, information on factor I_j is assumed to be available only for the initial and final years, instead of triennially. The PB approach is capable of exploiting the above extra information, so the final decomposition will be triennial (with partial information). Parameters θ_{Sj} and θ_y are fitted by GME estimation (Table 4.1). A summary is provided below (more detailed results appear in Fernández Vázquez and Fernández González 2008):

Tables 4.2–4.5 contain additive decomposition results for the change in greenhouse gas emissions in each of the settings considered: periodwise/unique (Sun's and LMDI approaches), triennial series with either partial (PB approach) or

---

[6] Countries included in the study are: Belgium, Denmark, Germany, Greece, Spain, France, Ireland, Italy, Luxembourg, the Netherlands, Austria, Portugal, Finland, Sweden and the United Kingdom.

Table 4.1 GME parameter estimates, PB approach

Country	(1) $\hat{\theta}_{I_j}$	(2) $\hat{\theta}_{S_j}$	(3) $\hat{\theta}_y$
Belgium	1.00	10^{20}	1.27
Denmark	1.00	0.684	1.27
Germany	1.00	10^{20}	1.27
Greece	1.00	0.907	1.27
Spain	1.00	10^{20}	1.27
France	1.00	3.618	1.27
Ireland	1.00	2.186	1.27
Italy	1.00	0.373	1.27
Luxembourg	1.00	0.819	1.27
Netherlands	1.00	0.681	1.27
Austria	1.00	0.298	1.27
Portugal	1.00	0.678	1.27
Finland	1.00	0.298	1.27
Sweden	1.00	0.414	1.27
United Kingdom	1.00	4.384	1.27

From Fernández Vázquez and Fernández González (2008)

Table 4.2 Additive decomposition of the variation in greenhouse gas emissions in the EU15 between years 1990 and 2002. PB method

Decomposition	Activity	Structural	Intensity	Total
Triennial (partial information)	2,233,440.00	74,227.90	−2,415,123.40	−107,455.44

Values of the effects in thousand tonnes of CO_2 equivalent
In the PB approach the effects are calculated by replacing the parameter estimates from Table 4.1 in the decomposition formulas (4.22−4.24)

complete information (Sun's and LMDI methods), and annual time series (Sun's, splines and LMDI approaches). Note that the decomposition is exhaustive in all methods applied, since the sum of the effects coincides with the total variation in emissions, although slight differences in the estimated effects are obtained depending on the specific approach and informative scenario considered.

Firstly, a significant reduction in greenhouse gas emissions (107,455.44 thousand tonnes of CO_2 equivalent) is observed in the EU15 along the study period. Regardless of the approach applied, that reduction is exclusively due to the strongly negative influence of the intensity effect. For instance, according to the annual decomposition provided by the splines method, adoption of cleaner technologies would have led, *ceteris paribus*, to a reduction of 2,389,401.46 thousand tonnes of CO_2 equivalent in greenhouse gas emissions for the whole EU15. On the contrary, the activity effect—and to a lesser extent also the structural effect—contributed to increase greenhouse gas emissions.

Table 4.3 Additive decomposition of the variation in greenhouse gas emissions in the EU15 between years 1990 and 2002. Sun's method

Decomposition	Activity	Structural	Intensity	Total
Periodwise	2,329,969.18	85,242.62	−2,522,667.24	−107,455.44
Triennial	2,193,012.68	98,584.99	−2,399,053.11	−107,455.44
Annual	2,188,296.01	95,476.80	−2,391,228.25	−107,455.44

Values of the effects in thousand tonnes of CO_2 equivalent

Table 4.4 Additive decomposition of the variation in greenhouse gas emissions in the EU15 between years 1990 and 2002. LMDI method

Decomposition	Activity	Structural	Intensity	Total
Periodwise	2,199,753.98	68,157.81	−2,375,367.23	−107,455.44
Triennial	2,182,121.14	93,207.82	−2,382,784.40	−107,455.44
Annual	2,186,471.68	94,207.14	−2,388,134.26	−107,455.44

Values of the effects in thousand tonnes of CO_2 equivalent

Table 4.5 Additive decomposition of the variation in greenhouse gas emissions in the EU15 between year 1990 and 2002. Splines method

Decomposition	Activity	Structural	Intensity	Total
Annual	2,187,882.08	94,063.95	−2,389,401.46	−107,455.44

Values of the effects in thousand tonnes of CO_2 equivalent

4.3.2 Comparison with Annual Time Series Decomposition Results

As mentioned above, a dynamic approach that considers intermediate periods would help alleviate the non-uniqueness problem. The potential disadvantages of dynamic decompositions include computational cost and the need of a large amount of information. In our study, annual data for the 1990−2002 period were obtained from Eurostat. Figures 4.3–4.5 include, respectively, the activity, structural and intensity effects calculated by using the various approaches (Sun, PB, LMDI, splines), under each of the three baseline scenarios described above.[7]

As expected a priori, Figs. 4.3–4.5 indicate that the techniques providing the most extreme results are *periodwise* decompositions (in particular, Sun's approach for the activity and intensity effects and the LMDI method for the structural effect).

Among the above (all of them exhaustive) methodologies, a specific technique may be regarded as 'most accurate' if its results are 'closest' to those obtained

---

[7] In the PB approach it suffices to have triennial information for some of the factors (partial additional information).

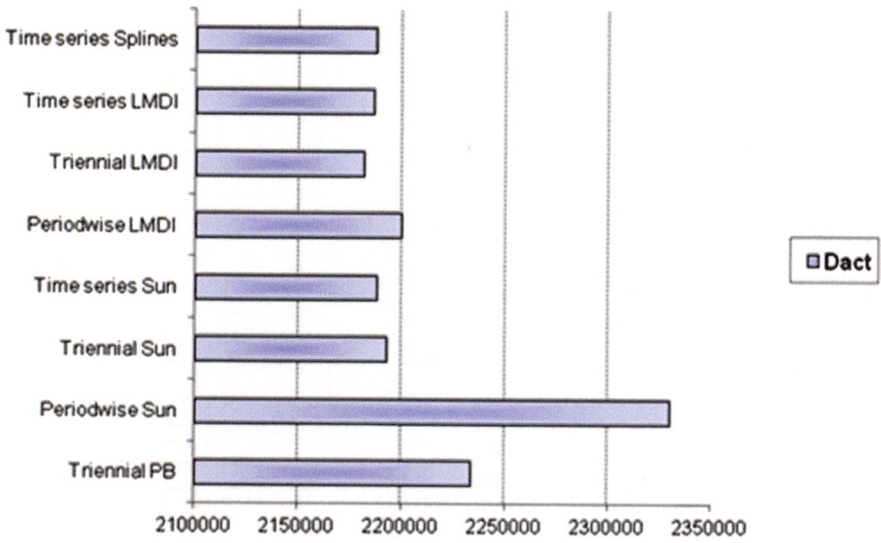

Fig. 4.3 Activity effect (EU15 1990–2002 period). Comparison of estimates obtained under the various approaches and information scenarios

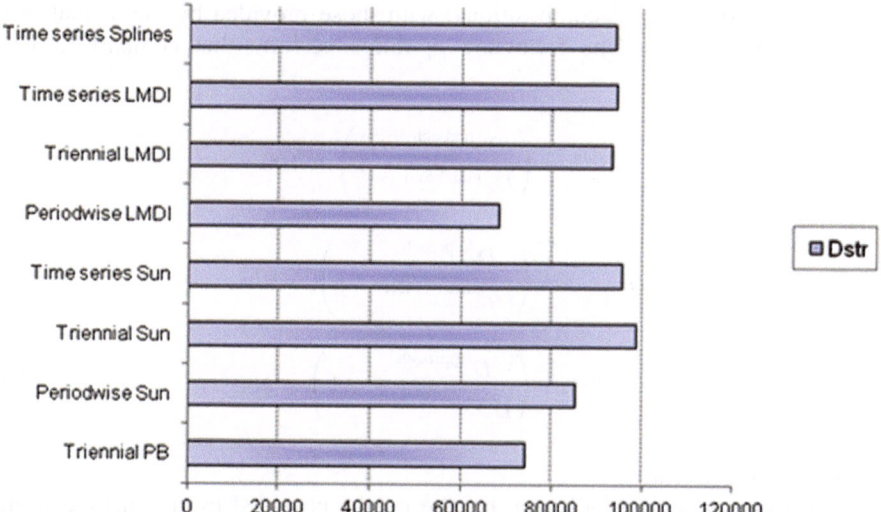

Fig. 4.4 Structural effect (EU15 1990–2002 period). Comparison of estimates obtained under the various approaches and information scenarios

in situations where the richest information is available (i.e. the annual time series decompositions carried out by using Sun, LMDI and splines). Tables 4.6–4.8 compare the effects estimated by applying the above approaches under several information situations (namely, periodwise and triennial—with either partial or

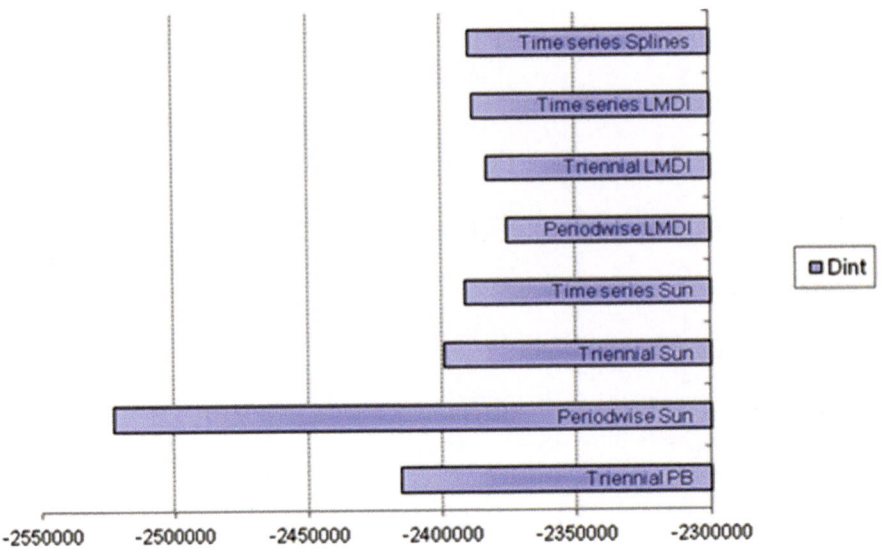

Fig. 4.5 Intensity effect (EU15 1990–2002 period). Comparison of estimates obtained under the various approaches and information scenarios

complete information—decompositions) with those provided by the annual time series decompositions. The following ρ_i ratios perform that comparison with relative terms:

$$\rho_i^{\text{Sun}} = \left(\frac{D_i^{\text{Approach}}}{D_i^{\text{Annual Sun}}} - 1 \right) \times 100 \tag{4.32}$$

$$\rho_i^{\text{LMDI}} = \left(\frac{D_i^{\text{Approach}}}{D_i^{\text{Annual LMDI}}} - 1 \right) \times 100 \tag{4.33}$$

$$\rho_i^{\text{Splines}} = \left(\frac{D_i^{\text{Approach}}}{D_i^{\text{Annual Splines}}} - 1 \right) \times 100 \tag{4.34}$$

where $i = $ act, str, int.

A ρ_i value near zero indicates that the results provided by the method in the numerator are very close to those of the baseline setting. In such cases, availability of annual data would not produce results that significantly differ from those obtained when only coarser (periodwise or triennial) information is at hand.

Overall, the results obtained from the three approaches are very similar for D_{act} and D_{int} effects. Only slight differences are observed for D_{str} when periodwise decomposition is chosen instead of any other alternative. In that case, the data available for the extreme years do not seem to adequately capture the structural

Table 4.6 Comparative analysis of the various decomposition methods *versus* annual Sun's approach

Approach	ρ_{act}^{Sun} (%)	ρ_{str}^{Sun} (%)	ρ_{int}^{Sun} (%)
Periodwise LMDI	0.524	−28.613	−0.663
Periodwise Sun	6.474	−10.719	5.497
Triennial PB	2.063	−22.256	0.999
Triennial Sun	0.216	3.255	0.327
Triennial LMDI	−0.282	−2.376	−0.353
Annual LMDI	−0.083	−1.330	−0.129
Annual splines	−0.019	−1.480	−0.076

Table 4.7 Comparative analysis of the various decomposition methods *versus* annual LMDI approach

Approach	ρ_{act}^{LMDI} (%)	ρ_{str}^{LMDI} (%)	ρ_{int}^{LMDI} (%)
Periodwise LMDI	0.607	−27.651	−0.535
Periodwise Sun	6.563	−9.516	5.633
Triennial PB	2.148	−21.208	1.130
Triennial Sun	0.299	4.647	0.457
Triennial LMDI	−0.199	−1.061	−0.224
Annual Sun	0.083	1.348	0.130
Annual splines	0.065	−0.152	0.053

Table 4.8 Comparative analysis of the various decomposition methods *versus* the annual splines approach

Approach	$\rho_{act}^{Splines}$ (%)	$\rho_{str}^{Splines}$ (%)	$\rho_{int}^{Splines}$ (%)
Periodwise LMDI	0.543	−27.541	−0.587
Periodwise Sun	6.494	−9.378	5.577
Triennial PB	2.082	−21.088	1.077
Triennial Sun	0.235	4.806	0.404
Triennial LMDI	−0.263	−0.910	−0.277
Annual Sun	0.019	1.502	0.076
Annual LMDI	−0.064	0.152	−0.053

changes occurring along the whole period. However, additional (triennial or annual) information seems to more properly collect some relevant changes experienced by that factor.

Tables 4.9–4.11 show the distances (absolute deviations, δ_i) and their sum (total deviation), obtained by comparing the effects calculated by the various methods (implemented under several informative scenarios), with those obtained in annual time series decompositions (the 'most informative' case) carried out through Sun's, LMDI and splines methods, respectively.

Table 4.9 Changes in greenhouse gas emissions (1990−2002)

Approach	δ_{act}	δ_{str}	δ_{int}	Total deviation
Periodwise LMDI	11,457.97	27,318.99	15,861.01	54,637.97
Periodwise Sun	141,673.17	10,234.18	131,438.99	283,346.35
Triennial PB	45,143.99	21,248.90	23,895.15	90,288.03
Triennial Sun	4,716.67	3,108.19	7,824.86	15,649.73
Triennial LMDI	6,174.87	2,268.98	8,443.85	16,887.70
Annual LMDI	1,824.33	1,269.66	3,093.99	6,187.98
Annual splines	413.94	1,412.85	1,826.79	3,653.58

Distance (in thousand tonnes of CO_2 equivalent) from the various approaches to annual time series decomposition implemented through Sun's method

Table 4.10 Changes in greenhouse gas emissions (1990–2002)

Approach	δ_{act}	δ_{str}	δ_{int}	Total deviation
Periodwise LMDI	13,282.31	26,049.33	12,767.02	52,098.66
Periodwise Sun	143,497.51	8,964.52	134,532.99	286,995.02
Triennial PB	46,968.32	19,979.24	26,989.14	93,936.70
Triennial Sun	6,541.00	4,377.85	10,918.86	21,837.71
Triennial LMDI	4,350.54	999.32	5,349.86	10,699.72
Annual Sun	1,824.33	1,269.66	3,093.99	6,187.98
Annual splines	1,410.40	143.19	1,267.20	2,820.79

Distance (in thousand tonnes of CO_2 equivalent) from the various approaches to annual time series decomposition implemented through LMDI

Tables 4.9–4.11 suggest a number of considerations. First, they clearly illustrate the ability of the PB approach to exploit additional information (being able to perform a triennial decomposition even with partial information). In periodwise decompositions, that advantage appears to lead to 'more accurate' results than direct application of Sun's method. On the other hand, when the comparison is made with periodwise LMDI, the latter seems to offer more accurate results in terms of absolute distances. In any event, the PB method emerges as a very useful alternative in regional and small-scale studies, where the lack of sufficient information for all factors often renders implementation of dynamic decompositions impossible.

Secondly, in the 'least informative' (periodwise decomposition) case, the LMDI approach looks remarkably 'more accurate' than Sun's method, regardless of the specific type (Sun, splines, LMDI) of annual time series decomposition method we used as a benchmark.

Third, in 'intermediate information' (triennial decomposition) situations, the 'most accurate' method depends on the specific decomposition technique that is used for the annual time series decomposition. If the comparison is made with Sun's method, the triennial decomposition with the same approach leads to 'more accurate' results (excepting the case of the structural effect). However, if the

Table 4.11 Changes in greenhouse gas emissions (1990–2002)

Approach	δ_{act}	δ_{str}	δ_{int}	Total deviation
Periodwise LMDI	11,871.91	25,906.13	14,034.22	51,812.26
Periodwise Sun	142,087.11	8,821.33	133,265.78	284,174.22
Triennial PB	45,557.92	19,836.05	25,721.94	91,115.91
Triennial Sun	5,130.61	4,521.05	9,651.65	19,303.31
Triennial LMDI	5,760.93	856.13	6,617.06	13,234.12
Annual Sun	413.94	1,412.85	1,826.79	3,653.58
Annual LMDI	1,410.40	143.19	1,267.20	2,820.79

Distance (in thousand tonnes of CO_2 equivalent) from the various approaches to annual time series decomposition through the splines method

reference approach is LMDI or splines, the LMDI method would be the 'most accurate' one.

Finally, in the 'full information' (time series decomposition) setting the results obtained through LMDI and splines are similar and relatively differ from those provided by Sun's method. More specifically, the splines method has the smallest absolute distance regardless of the reference approach. Therefore, in that scenario, the splines method seems to be the most accurate one among the three techniques.

Evidently, the scope of the above conclusions is limited to our dataset. More extensive analysis (including studies on other databases, as well as simulation studies) would be required in order to test for general validity of these qualitative findings.

4.3.3 A Cross-Country Analysis

A detailed study on the variation of greenhouse gas emissions in each EU member state is very useful in order to complete overall analysis. To that end, we shall additively decompose the aggregate by using the splines method as—according to the above analysis—it seems to be the most accurate approach. A summary of decomposition results at the country level is collected in Table 4.12.

It is observed in Table 4.12 that the reducing influence of the intensity effect is brought about mainly by large economies such as Germany, the United Kingdom and even France. The sum of German and British contributions is more than half of the total amount of the intensity effect. Italy, Spain and the Netherlands also have significant shares.

Secondly, the activity effect is positive and of large magnitude, thus significantly contributing to increase emissions and to absorb the reducing boost of the intensity effect. All the countries studied—especially large economies such as Germany, the United Kingdom, France, Italy and Spain, plus the Netherlands—withstood more significant pressures on their respective emission levels as a result of economic growth.

Table 4.12 Results of additive (annual time series) decomposition by country, 1990–2002 period. Splines method

Countries	D_{act}	D_{str}	D_{int}	D_{tot}
Belgium	79,768.98	−1,852.02	−73,659.60	4,257.36
Denmark	40,464.11	1,427.15	−42,151.21	−259.95
Germany	578,771.02	−50,202.93	−761,174.44	−232,606.35
Greece	61,791.63	27,096.37	−58,290.80	30,597.20
Spain	174,556.06	16,550.99	−75,930.28	115,176.77
France	300,076.47	−35,206.76	−275,714.54	−10,844.83
Ireland	32,193.53	44,601.04	−61,339.43	15,455.14
Italy	276,684.87	−76,675.16	−155,154.84	44,854.87
Luxembourg	5,625.63	4,855.30	−12,401.31	−1,920.38
Netherlands	115,911.08	26,376.81	−139,906.80	2,381.09
Austria	42,651.15	−601.13	−35,176.93	6,873.09
Portugal	36,332.85	19,546.45	−32,171.45	23,707.85
Finland	41,023.36	−18,555.59	−17,274.88	5,192.89
Sweden	38,130.51	−15,415.00	−25,254.40	−2,538.89
UnitedKingdom	363,900.83	152,118.42	−623,800.55	−107,781.30
EU15	2,187,882.08	94,063.95	−2,389,401.46	−107,455.44

In thousand tonnes of CO_2 equivalent

Finally, the structural effect was slightly positive for the whole EU15, increasing the level of pollutant emissions by a smaller amount than the activity effect. However, the country-level analysis (Table 4.12) reveals that in some member states—such as Finland, Sweden, Belgium, Austria, and especially in large economies such as Italy, Germany and France—the structural change induced a reduction in the emission levels. On the contrary, in others countries such as Denmark, Luxembourg, Spain, Portugal, the Netherlands, Greece, Ireland and the United Kingdom, the opposite effect occurred: in the latter two countries, the upward pressures on the emission levels were so strong that they alone were able to offset the lowering influences coming from other EU15 countries.

Overall, the level of greenhouse gas emissions in the EU15 decreased by 107,455.44 thousand metric tonnes of CO_2 equivalent. Despite the pressures caused by factors such as GDP growth at the whole EU15 level and—to a lesser extent—structural changes, other actions—including the promotion of innovation and adaptation to more efficient technologies, as well as R&D, technical change and replacement of fossil fuels by less polluting energy sources—have served as effective counterbalances in order to reach the goal of reducing gas emissions.

4.4 Conclusions

A well-known practical problem in decomposition analysis is that, despite the numerous formulas being equivalent from a theoretical standpoint, results often depend heavily on the chosen decomposition formula. Sun (1998) proposed a solution to the non-uniqueness problem: the 'jointly created and equally distributed' principle. Afterwards, Fernández Vázquez (2004) derived the path-based method—which may be seen as a generalization of Sun's approach—and subsequently applied it to index-based decomposition analysis in Fernández Vázquez and Fernández González (2008). The solutions provided by the PB method depend on a set of unknown parameters characterizing the optimal trajectories of the determinant factors. Generalized maximum entropy econometrics is used to estimate those parameters, so the decomposition formula that provides optimal fit to the additional information is chosen. The method is useful when additional, possibly very limited information is available for some factors in intermediate periods.

In this chapter, we have applied the above two methods, together with the LMDI and splines approaches, in order to quantify the contributions of three factors (namely, GDP of the EU15, share of national GDP in global EU15 GDP, and greenhouse gas emissions per unit of national GDP) to the change in greenhouse gas emissions in the EU15 along the 1990–2002 period. The decomposition was carried out under several initial information scenarios. Regardless of the implemented methodology and baseline information, our findings show a large reduction in greenhouse gas emissions as a consequence of the intensity effect. By contrast, the activity and—to lesser extent—structural effects helped increase emissions. Since we expect economic growth to continue in the years to come—and given that evidence suggests we cannot rely on structural change as a reducing force—a number of action lines—including use of higher quality energies, technical innovation and technological change—will be indispensable in reducing emissions.

We have also included in this chapter a comparison analysis on the performance of the various decomposition methods in our specific setting. Results suggest that the ability of the PB approach to exploit partial additional information can lead to 'more accurate' results that significantly differ from those provided by other methods (such as Sun's approach) that are less demanding in terms of information availability. Thus, the PB approach is of particular interest in regional studies, where lack of information on some factors often prevents dynamic decompositions from being applied. However, in spite of all these advantages, our case analysis found no evidence that the PB approach (relying on partial, triennial information) produces 'more accurate' results than those provided by a simpler, periodwise LMDI decomposition. Comparisons also indicate that in 'poor information' scenarios, application of LMDI method would lead to 'more accurate' results than Sun's approach. In addition, when 'complete' (annual) information is available, the splines method appears to deliver more accurate results than LMDI and Sun's approach. Overall, the splines and LMDI methods seem to offer more robust results than the other methods considered in this chapter.

References

Albrecht J, Francois D, Schoors K (2002) A shapley decomposition of carbon emissions without residuals. Energy Policy 30(9):727–736

Ang BW (2004) Decomposition analysis for policymaking in energy: which is the preferred method? Energy Policy 32(9):1131–1139

Ang BW, Choi KH (1997) Decomposition of aggregate energy and gas emission intensities for industry: a refined Divisia index method. Energy J 18(3):59–73

Ang BW, Lee SY (1994) Decomposition of industrial energy consumption: some methodological and application issues. Energy Econ 16(2):83–92

Dietzenbacher E, Los B (1998) Structural decomposition techniques: sense and sensitivity. Econ Syst Res 10(4):307–323

Eurostat: New Cronos data base, Statistical Office of the European Communities Regions, Luxembourg

Fernández González P (2013) Técnicas de descomposición de variaciones basadas en indices Divisia. Algunas aplicaciones medioambientales. Thesis, Universidad de Oviedo (Spain)

Fernández Vázquez E (2004) The use of entropy econometrics in decomposing structural change. Thesis, Universidad de Oviedo (Spain)

Fernández Vázquez E, Fernández González P (2008) An extension to Sun's decomposition methodology: the path based approach. Energy Econ 30(3):1020–1036

Golan A, Judge G, Miller D (1996) Maximum entropy econometrics: robust estimation with limited data. John Wiley and Sons, Chichester (UK)

Harrison WJ, Horridge JM, Pearson KR (2000) Decomposing simulation results with respect to exogenous shocks. Comput Econ 15(3):227–249

Sun JW (1998) Changes in energy consumption and energy intensity: a complete decomposition model. Energy Econ 20(19):85–100

Vogt A (1978) Divisia indices on different paths. In: Eichhorn W et al (eds) Theory and application of economic indices. Physica-Verlag, Wurzburg

Concluding Remarks

Energy is essential in everyday life. For this reason, it is nowadays crucial to address major energy challenges such as the climate change, the increasing dependence on imports, the pressure on energy resources and the supply to all consumers of secure, affordable energy. Numerous publications discuss the economic developments that are threatening to jeopardize natural resources and the environment, preventing or hindering sustainable growth.

In this book we have tried to address the goals of identifying, quantifying and analysing the impact of the specific factors influencing some relevant environmental and energy aggregates such as energy intensity and greenhouse gas emissions. We have focused on a Divisia-index-based methodology that allows researchers to decompose the variation of a generic aggregate—absolute value, ratio or elasticity, among other possibilities—into a set of predefined factors, with the relevant variables being assigned weights that are allowed to evolve in time according to any suitable criterion selected by the researcher. This methodology is widely accepted in the fields of energy and environment and—despite its relative complexity and the need for the researcher to select weights—it has the advantage of not requiring large amounts of initial information, also allowing a simple interpretation of results.

We have structured the work in two parts. In the first one (Chaps. 1 and 2), we have focused on theory. In Chap. 1, we reviewed the literature on Divisia-index-based techniques to decompose variations. We focused on exhaustive techniques, paying especial attention to the LMDI method. This is currently an active research field, as researchers continue working on devising improved techniques to analyse the driving forces underlying the changes in environmental indicators. In this regard, recent works aim at developing new decomposition approaches to quantify sectoral contributions to changes in an aggregate. Among these new techniques, the methods of *dematerialization* (Ang and Xu 2013) and *attribution analysis* (Choi and Ang 2012; Fernández González et al. 2013) appear particularly promising. On another front, decomposition of changes in aggregates related to renewable resources and CO_2 emissions also announces some interesting challenges for researchers.

In Chap. 2 we have introduced a new decomposition method that provides a nonparametric alternative to standard decomposition techniques, allowing to

P. Fernández González et al., *The Driving Forces of Change in Environmental Indicators*, Lecture Notes in Energy 25, DOI: 10.1007/978-3-319-07506-8, © Springer International Publishing Switzerland 2014

exhaustively decompose the variation in a generic aggregate. The method uses classical spline interpolation in order to obtain approximations for the theoretical effects derived in the continuous time model. Convergence of the spline-based approximations to these theoretical quantities is also derived in the chapter. The splines method may be used to decompose variations of many aggregates, including absolute values, energy intensities and elasticities, as well as sums and products of any finite number of components or time paths. More generally, it may be applied to decompose the variation of an arbitrary differentiable functional of a vector of smooth trajectories. As an interesting research goal, the splines method can also be extended to include information from other variables/signals, with a view to improving the trajectory reconstruction process, and thus the accuracy of the approximations for the continuous time effects. This may be achieved—among other possibilities—by exploiting results from the theory of dynamic splines (e.g. Kano et al. 2003; Sun et al. 2000).

In the second part (Chaps. 3 and 4), we carried out a number of applications that analyse variations in some specific aggregates. In Chap. 3, we multiplicatively decomposed the change in aggregate energy intensity in the European Union between years 1995 and 2010. We relied on the splines method proposed in Chap. 2, as well as in some classical—both parametric and nonparametric—Divisia-index-based techniques. The various methods provided similar results. In particular, the splines approach delivered an exhaustive decomposition with results fairly close to those obtained from the LMDI method. During the study period, aggregate energy intensity decreased by about 13 % as a result of the intensity factor. On the other hand, structural changes in the EU contributed to increase the aggregate, partly offsetting the reducing influence of the intensity effect. We also included energy price—more specifically, oil price—as a relevant variable in the analysis. Its evolution exhibits a growing trend having a strongly negative correlation with the downward trajectory of the intensity and total effects. This would suggest that energy price increases may stimulate investment in more efficient technologies, technical change, innovation and the use of higher quality energy.

In Chap. 4, we additively decomposed the change in greenhouse gas emissions in the EU15 for the 1990–2002 period. The specific factors considered were the GDP of the EU15 (*activity effect*), the share of national GDP in the GDP of the EU15 (*structural effect*) and greenhouse gas emissions per unit of national GDP (*intensity effect*). We applied several exhaustive decomposition methodologies, including the LMDI and splines methods, as well as other additive approaches (namely Sun's and PB methods). Several potential scenarios were considered, each with a different amount of information (respectively, periodwise, triennial and annual) available. Our analysis suggests that the advantage offered by the PB method of incorporating partial triennial information may produce 'more accurate' results than periodwise Sun's method, although the same evidence also indicates that (both periodwise and triennial time series) LMDI decompositions lead to 'more accurate' (in the sense of being closer to those offered by annual time series decompositions) results than Sun's and PB approaches. Finally, implementation of the splines method delivered results similar to those obtained from the other methods, particularly the LMDI method based on annual time series. This emphasizes the role of the splines method as a workable alternative to mainstream Divisia-index-based decomposition methods, including the heavily used LMDI approach. The results obtained in the chapter indicate a reduction of greenhouse gas emissions in Europe by around 107 million tonnes of CO_2 equivalent, with the intensity effect improving emissions efficiency, whereas economic growth and structural change pushed up the aggregate.

A detailed analysis of the results from Chaps. 3 and 4 allows us to derive some general conclusions that may be useful in order to devise useful strategies and a number of energy/environmental performance lines for the studied countries. In particular, with a view to pursuing sustainable development, the following policies may be recommended: (a) promotion of structural change towards less energy-intensive and higher added value goods and services, (b) establishment of energy efficiency measures (especially in those sectors—such as steel, cement, power, chemicals and transport—that are intensive in energy and/or emissions), (c) research and development in energy-saving processes, (d) adoption of more efficient and less polluting equipment, (e) implementation of technologies for scavenging of gaseous pollutants, (f) development of cogeneration and trigeneration processes, (g) reduction in the use of fossil fuels and (h) promotion of 'green attitudes' among the population. All these action lines reveal themselves as necessary to offset the potentially adverse impacts of other factors mentioned in our analysis.

The International Energy Agency, in its *Energy Technology Perspectives* 2012, formulates a number of specific policy recommendations for the above challenges that are consistent with the results obtained in this work. Its high-level recommendations to the energy ministers include:

(i) Creation of an environment conducive to investment, enabling the launch of clean energy. Industry is essential for that transition. The existence of common goals backed by rigorous and predictable policies is essential to establish the necessary credibility within the investment community.

(ii) Paving the way for clean energy technologies. Governments should commit themselves to advance national activities aimed at adequately reflect the true cost of production and consumption, and to report such actions. In this regard, some proposals are intended to put a price on CO_2 emissions and to phase out inefficient fossil fuel subsidies, while ensuring access to affordable energy for all citizens.

(iii) Increase in the efforts to unlock the potential of energy efficiency, looking for energy efficiency improvements in all energy consumer sectors.

(iv) Acceleration of innovation and research, as well as development and education of the population on energy issues. Public authorities should develop and implement strategic plans for energy research, stimulated by a major and sustained financial support. In addition, governments should consider joint efforts in R&D in order to coordinate their activities and avoid duplication, in addition to improving outcomes, reducing the cost of technology to early-phase innovation and sharing lessons learned on models of innovation in R&D&I.

An integrated use of existing key technologies would make possible a reduction of the dependence on imported fossil fuels and limited national resources as well as enabling low carbon content electricity, also improving energy efficiency and reducing emissions in the industry, transport and building sectors. This would also slow down the rapid growth in the demand for energy, thus reducing imports, strengthening the national economies and, over time, reducing greenhouse gas emissions.

Extra Contents and Instructions

The extra materials accompanying the text include the following files:

- 'readme.txt': this is a help file in plain text format.
- 'Chapter3.xls' and 'Chapter4.xls': these are Excel files containing the data sets and algorithms used in implementing the decomposition techniques (excepting the splines method) applied in Chaps. 3 and 4, respectively.
- 'datachapter3.csv' and 'datachapter4.csv' are the same data sets, in csv format files. These are the inputs for the spline-based analysis in Chaps. 3 and 4.
- 'dataread.m' is a Matlab m-function that reads the databases in csv format.
- 'chapter3splines.m' is a Matlab executable file ('m-file') that implements the splines method as used in Chap. 3.
- 'chapter4splines.m' is an m-file for the spline-based analysis required in Chap. 4.

Note: The above m-files require the Spline Toolbox of Matlab installed on your computer.

P. Fernández González et al., *The Driving Forces of Change in Environmental Indicators*, Lecture Notes in Energy 25, DOI: 10.1007/978-3-319-07506-8, © Springer International Publishing Switzerland 2014

Specific Instructions for Using the MATLAB Codes

First, copy 'data chapter3.csv' and 'datachapter4.csv' to the desired directory (by default, the root directory C:). Then copy 'dataread.m', 'chapter3splines.m' and 'chapter4splines.m' to the desired location (e.g. the 'bin' folder of Matlab), making sure that the location is accessible from Matlab (the proper 'path' must be specified, if necessary). The files 'chapter3splines.m' and 'chapter4splines.m' can be opened and edited by using Matlab Editor. A different data set may be specified (provided the information it contains is structured as in datachapter3.csv or in datachapter4.csv). Both multiplicative and additive decompositions are allowed. By typing the function name (chapter3splines or chapter4splines, respectively) on Matlab's command window, the selected function is executed and the results of the spline-based decomposition analysis are displayed on screen.

References

Ang BW, Xu XY (2013) Tracking industrial energy efficiency trend using index decomposition analysis. Energy Econ 40:1014–1021

Choi KH, Ang BW (2012) Attribution of changes in Divisia real energy intensity index: an extension to index decomposition analysis. Energy Econ 34(1):171–176.

European Commission. Communication from the Commission—Energy efficiency: delivering the 20 % target, COM(2008) 772. http://eur-lex.europa.eu/LexUriServ/LexUriServ.do?uri=CELEX:52008DC0772:EN:NOT. Accessed 13 Nov 2008

European Commission (2011) European Economic Statistics. Publications. Office of the European Union: Luxembourg. http://epp.eurostat.ec.europa.eu/portal/page/portal/statistics/searchdatabase

European Commission (2013) Statistical Office of the European Union, Luxembourg. http://epp.eurostat.ec.europa.eu/portal/page/portal/statistics/themes

Eurostat: New Cronos data base, Statistical Office of the European Communities Regions, Luxembourg.

Fernández González P, Landajo M, Presno MJ (2013) The Divisia real energy intensity indices: evolution and attribution of percent changes in 20 European countries from 1995 to 2010. Energy 58(1):340–349

Fernández González P, Landajo M, Presno MJ (2014) The driving forces behind changes in CO_2 emission levels in EU-27. Differences between member states. Environ Sci Policy 38:11–16

P. Fernández González et al., *The Driving Forces of Change in Environmental Indicators*, Lecture Notes in Energy 25, DOI: 10.1007/978-3-319-07506-8, © Springer International Publishing Switzerland 2014

Fernández González P, Pérez Suárez R (2000) Descomposición de la intensidad de energía agregada industrial en España: el efecto estructural y el efecto intensidad. XIV Reunión ASEPELT, Oviedo (SPAIN)

Fraser I (2000) An application of maximum entropy estimation: the demand for meat in the United Kingdom. Appl Econ 32(1):45–59

Gardebroek C, Lansink A (2004) Farm-specific adjustment costs in Dutch pig farming. J Agric Econ 55(1):3–24

Gardner DT, Elkhafif MAT (1998) Understanding industrial energy use: structural and energy intensity changes in Ontario industry. Energy Econ 20(1):29–41

Gardner DT, Robinson JB (1993) To what end? A conceptual framework for the analysis of energy use. Energy Stud Rev 5(1):1–14

Golan A, Judge G, Miller D (1996) Maximum entropy econometrics: robust estimation with limited data. John Wiley and Sons, Chichester

Golan A, Perloff JM, Shen EZ (2001) Estimating a demand system with nonnegativity constraints: Mexican meat demand. Rev Econ Stat 83(3):541–550

Hankinson GA, Rhys MM (1983) Electricity consumption, electricity intensity and industrial structure. Energy Econ 5(3):146–152

International Energy Agency (2008) World energy outlook. OECD, Paris. http://www.iea.org/publications/freepublications/publication/name,26707,en.html. Accessed 2011

International Energy Agency (2010) Key world energy statistics. Soregraph, France

International Energy Agency (2012) Energy technology perspectives 2012. IEA Publications, France

Jollands N, Aulakh HS (1996) Energy use patterns and energy efficiency trends: the case of energy intensity analysis in New Zealand. IPENZ Annual Conference 1996, Proceedings of engineering, providing the foundations for society, vol 2, pp 95–100. http://www.ema.org.nz/papers/96eupeet.htm

Kano H, Egerstedt MB, Nakata H, Martin CF (2003) B-splines and control theory. Appl Math Comput 145(2):265–288

Kapur JN, Kesavan HK (1993) Entropy optimization principles with applications. Academic Press, New York

Nathwani JS, Siddall E, Lind NC (1992) Energy for 300 years. Institute for Risk Research, University of Waterloo, Ontario, Canada

Paris Q, Howitt RE (1998) An analysis of Ill-posed production problems using maximum-entropy. Am J Agric Econ 80(1):124–138

Ross M, Larson ED, Williams RH (1987) Energy demand and material flows in the economy. Energy 12(10–11):1111–1120

Sun S, Egerstedt MB, Martin CF (2000) Control theoretic smoothing splines. IEEE Trans Automat Control 45(12):2271–2279

United Nations Statistics Division (Environment and Energy Statistics). http://unstats.un.org/unsd/environment_main.htm

United Nations (1992) United Nations Framework Convention on Climate Change. United Nations

United Nations (1997) Kyoto Protocol to the United Nations framework convention to the climate change. United Nations

Vartia YO (1974) Relative changes and economic indices. Licentiate Thesis, Department of Statistics, University of Helsinki

Williams RH, Larson ED, Ross MH (1987) Materials, affluence and industrial energy use. Annu Rev Energy 12:99–144